Science as a Human Endeavor

Science as a Human Endeavor

George F. Kneller

Columbia University Press

New York   1978

Library of Congress Cataloging in Publication Data

Kneller, George F., 1908–
    Science as a human endeavor.

    Bibliography: p.
    Includes index.
    1. Science. 2. Science—Philosophy. 3. Science—
Social aspects. I. Title.
Q172.K66   500   77-19167
ISBN 0-231-04206-X

Columbia University Press
New York and Guildford, Surrey

And I gave my heart to seek and search out by wisdom
concerning all things that are done under Heaven:
this sore travail hath God given to the sons of man
to be exercised therewith.

Ecclesiastes I;13

Preface

Science has always been controversial. It has been welcomed by some for its commitment to the rational solution of problems and to the advance of testable knowledge. It has been rejected by others for its opposition to traditional thought and its attack on mysticism. Today it is defended by those who prize the high standard of living that science makes possible. It is criticized by others who claim that it is misdirected by the interests of its clients, or that it is a self-moving force indifferent to human concerns.

Why does science give rise to such conflicting views? As a human undertaking, science is fallible; it can degenerate or it can respond to men's highest aspirations. As a part of society, science also is open to outside influences; like any social enterprise, it can be used or misused. Thus different aspects of science arouse different responses. In this book, however, I seek to correct these partial responses by portraying science in its entirety. I inquire into its powers and limitations, into the threats it poses and the promises it holds. Throughout I seek to show that science is a human endeavor and not an impersonal juggernaut.

This work is intended for scientists and humanists alike. It should help scientists to see the relevance and interdependence of their specialties, and humanists to understand what scientists are trying to do. It presupposes no more scientific knowledge than what is contained in a good basic course in general science. Although I have had the lay reader in mind throughout, I have made no attempt to popularize my subject matter. I have clarified terms, simplified concepts, and provided many illustrations, but I have not tried to make science appear either entertaining or easy. Science is an enormously complex enterprise that brings the intellect to its peak, and the lay reader must be willing to pause and reflect on his reading. If he does so, he will be richly rewarded, for science is as fascinating and challenging as any human quest.

No one realizes more keenly than I the risk I have run in seeking to cover so much so briefly. Specialists in every topic I treat will know more about it than I do. Yet if books of any scope are to be written, depth in the specialty must be sacrificed to the coherence of the whole. Some experts will maintain that a work on this subject should be entrusted to a number of specialists, each treating an aspect. Yet that approach does not exclude mine. For no set of contributors can form a single point of view, and only the single point of view can recreate for the reader the unity of the scientific enterprise in all its spheres.

This does not mean that the single writer must do his work alone. I have profited immensely from the criticisms of the many specialists who have read individual chapters, as well as from appraisals of the work as a whole by Professor Robert M. Westman and the Columbia University Press reviewers. Without the help of John M. Harrison the book might never have appeared. I am grateful to all these persons for their valuable advice. Nevertheless, what appears here is the product of a single mind aware of its limitations. If I have done less than justice to any topic I have treated, I hope that this will be balanced against my attempt to do justice to the whole.

George F. Kneller
University of California
Los Angeles

Contents

Science as a Human Endeavor

Chapter 1

Science in History

Simply put, science is knowledge of nature and the pursuit of that knowledge. Yet this pursuit involves a great deal. It involves, among other things, a history, a method of inquiry, and a community of inquirers. Today especially, science is a cultural force of overwhelming importance and a source of information indispensable to technology. My aim in this book is to explain these aspects of science and show how they are interrelated.

SCIENCE AND THE ORDER OF NATURE

Glancing through history, we find that nature has been studied for a variety of reasons. In Aristotle's Lyceum[1] it was studied to enlighten and improve the seeker of knowledge; in Renaissance Europe, to display God's design in His creation; in modern times, to advance knowledge both for its own sake and for its social and technical uses. But these grand purposes seem to have inspired scientists less than two primal emotions—wonder and fear. Early man was largely at the mercy of nature. Perhaps his strongest motive for natural inquiry was to attain peace of mind through having some plausible explanation of natural disasters. He wanted to find out what caused earthquakes, floods, fire, and disease. In China the Taoist natural philosophers, in ancient Europe the Stoics, the Epicureans, and the followers of the atomist Democritus, all practiced science from this motive.[2] Epicurus wrote that "if we were not troubled at all by apprehensions about phenomena in the sky and concerning death, lest it somehow concern us, and again by our failure to perceive the limits of pains and desires, we should have no need of the study of nature."[3]

Fear is allayed by the recognition that nature is orderly and in-

1

telligible. Wonder begins with this recognition. As science grew and men began to master the world, wonder became the driving force behind the greatest scientific achievements. Einstein made this point eloquently:

> the cosmic religious feeling is the strongest and oldest motive for scientific research. Only those who realize the immense efforts and, above all, the devotion without which pioneer work in theoretical science cannot be achieved are able to grasp the strength of the emotion out of which alone such work, remote as it is from the immediate realities of life, can issue. What a deep conviction of the rationality of the universe and what a yearning to understand it . . . Kepler and Newton must have had to enable them to spend years of solitary labor in disentangling the principles of celestial mechanics![4]

What, then, is the "order of nature"? For many people it is the laws of the heavens, admired by the philosopher Kant, who compared them to the "laws in our breasts," and celebrated by George Meredith in his poem "Lucifer in Starlight":

He reached a middle height, and at the stars,
Which are the brain of heaven, he looked, and sank.
Around the ancient track marched, rank on rank,
The army of unalterable law.[5]

To the scientist, however, all of nature is interrelated and, as such, orderly. Instead of being a chaos, the universe is a single grand nexus of things and processes. No event, he holds, is utterly unconnected with others and hence inexplicable. Whatever seems unconnected will, with continuing inquiry, be found to occur only in conjunction with other events. So-called freak events—hurricanes, plagues, explosions of galaxies—are as orderly in this sense as the wheeling of the planets and the ripening of corn.

Thus the order of nature is whatever remains invariant among the changes of things and is the cause of those changes. These invariant features of nature are fixed patterns in events at all levels from atoms to galaxies. The fall of the apple on Newton's head in his fam-

2

ily's garden at Woolsthorpe could not in practice have been predicted. Yet it was orderly, for it obeyed the same gravitational force that keeps the stars in their courses and the tides to their ebb and flow.

The aim of science is to reach an exact and comprehensive understanding of the order of nature. Because the constituents of nature are almost infinitely diverse, this quest has taken many centuries and will take many more. Hence science is intrinsically historical. Not only scientific knowledge but the techniques by which it is produced, the research traditions that produce it, and the institutions that support them all change in response to developments among themselves and in the social and cultural world to which they belong. If we are to understand what science really is, we must regard it first and foremost as a succession of movements within the greater historical movement of civilization itself.

OTHER CIVILIZATIONS, OTHER SCIENCES

I say a "succession of movements" because history reveals not one science but several. In every civilization certain men have thought systematically about the natural world and have sought the causes of phenomenal change in nature itself rather than in human or suprahuman volition. But until the Arabs inherited Greek natural philosophy and Chinese alchemy and transmitted them to the West, there was no single body of natural knowledge that passed from one civilization to another. On the contrary, in every civilization the study of nature took its own path. Greek and Chinese natural philosophers explained much the same physical world very differently. The Greeks proposed the theory of the four elements (earth, air, fire, water) and the theory that everything in the universe has its natural place. The Chinese used the theory of opposing natural forces, yin and yang, and the theory of the five phases through which all things pass in cycles. We call these different cultural traditions "science" not because they form a single historically evolving entity, but because they are different historical entities of the same general kind.

But this judgment depends on hindsight. In China, classical Greece, Islam, and medieval Europe there was no term equivalent to our "science" and there was no scientific community. The activities we group together as Greek or Chinese science were carried out by philosophers, mathematicians, astronomers, physicians, and others holding quite different views about the kind of inquiry they were pursuing. It is we who see in their work the characteristics of a science that they themselves could not recognize.

The latest of these scientific traditions, the Western or European, has proved astonishingly successful and (we think) has come closest to representing what nature is really like. Whereas previous sciences were culture bound, expressed in the language of a particular people, European science has become international and universal, for it is expressed in the supracultural language of mathematics and is practiced the world over.[6] Nevertheless, this science was not created by Europeans alone. Through a range of contacts—conquest, trade, diplomacy, travel—Europeans drew on the scientific and technological achievements of other civilizations. From the Greeks they inherited Ptolemaic astronomy, Euclidean geometry, Galenic medicine, the mathematical tradition of Plato and Pythagoras, and the more empirical tradition of Aristotle. From China came magnetic physics, explosives chemistry, astronomical coordinates, the idea of infinite space, quantitative cartography, and a stream of technological inventions such as gunpowder, paper, horse harnesses, the driving belt, the chain drive, and the sternpost rudder.[7] From India there came numerals, zero, algebra, a theory of atomism, and a rich pharmacology of herbs and minerals.

Most of these achievements were first absorbed by Islam, which from 750 A.D. to the late Middle Ages stretched from Spain to Turkestan. The Arabs unified this vast body of knowledge and added to it. They improved algebra, invented trigonometry, and built astronomical observatories. They invented the lens and founded the study of optics, maintaining that light rays issue from the object seen rather than from the eye. In the tenth century Alhazen discovered a number of optical laws, for example, that a light ray takes the quickest and easiest path, a forerunner of Fermat's "least action" principle.[8] The Arabs also extended alchemy, improving and inventing a wealth of techniques and instruments, such as the alembic, used to distill perfumes. In the eighth century the physician al-Razi laid the foundations of chemistry by organizing alchemical knowledge and denying its arcane significance. Inventor of the animal-vegetable-mineral classification, he categorized a host of substances and chemical operations, some of which, such as distillation and crystallization, are used today. When Arabic science declined, of the three great civilizations on the borders of Islam—China, India, and Europe—the last inherited its great synthesis.

In 1000 A.D. Europe was so backward that it had to borrow the Islamic sciences wholesale, translating Arabic writings into Latin. By 1400 European science had no superior. What caused this dramatic transformation? Why did modern science begin its exponential rise among the warring states of crowded Europe rather than in some older, more harmonious civilization? Why not in China, for instance?[9]

Chinese Science. As Joseph Needham has shown,[10] during the first fifteen centuries of the Christian era Chinese science was the equal of any and Chinese technology was probably superior to all. Certain sciences—astronomy, mathematics, hydraulic engineering—were supported by the state bureaucracy, which was imbued with the teachings of Confucianism. This philosophy studies man as a social being and proposes principles for the wise management of society. Other sciences—alchemy, biology, medicine (to some extent), physics (except for harmonics), and geology—remained unorthodox and were practiced largely by the Taoists, who studied man's inner life and his relation to nature. (Man, they said, should renounce ambition and live in accord with the order, or tao, of natural events.) Taoists inspired most of Chinese science, but they distrusted reason and speculation, while the Confucians were interested in science only for its social uses. As a result Chinese science tended to avoid theory and remained largely empirical.

Nevertheless, this empiricism was anything but crude. The Chinese observed and recorded accurately and persistently. Their astronomers noted the positions of stars and other celestial phenomena in measured degrees. Indeed, their lists of novae,[11] comets, and meteors are used by radio astronomers today. In the first century B.C.

5

their hydraulic engineers were recording the silt content of rivers precisely. To improve observation, the Chinese invented instruments such as the seismograph, the mechanical clock, and the magnetic compass. They classified many phenomena, such as stars, diseases, and medicinal herbs and minerals. They also carried out experiments; for instance, they tested the acoustical properties of bells and strings, and the strengths of different materials.[12]

What sciences developed with these methods? The Chinese had algebra but little geometry. Hence their theoretical astronomy remained weak. Unlike Greek geometry, which represented the movements of the heavenly bodies in three-dimensional space, Chinese algebraic techniques implied no particular physical hypothesis. Hence despite voluminous records they lacked an adequate theory of the heavens. In physics they had little mechanics and no dynamics, but they pioneered the science of magnetism and made an exhaustive study of their own music. During the Middle Ages their maps were much more accurate than European ones. In medicine they developed a comprehensive account of relations between body, mind, and environment. Their alchemy, the oldest in the world, sought the health-giving elixir of eternal life, an idea which did not appear in Europe until the twelfth century by way of Islam.

The Chinese had theories, too, but these were general and qualitative. According to the "two-force" theory, the fundamental forces in the universe are yin (expressed, for instance, in rain and femaleness) and yang (expressed in heat and maleness). The "five-phases" theory sought to classify the basic processes at work in nature, naming them water, fire, wood, metal, and earth. The phases supersede one another in cycles: wood supersedes earth; metal, wood; fire, metal; water, fire; earth, water; and then the cycle begins again. With these phases they correlated everything in the universe that could be classified in fives—tastes, smells, seasons, cardinal points, musical notes, planets, weathers, and so on. The fivefold correlations display the mutual affinities between things. All things in the same class (e.g., east, wood, green, wind, wheat) resonate with one another, exchanging energies. In the words of the philosopher Tung Chung-Shu, writing in the second century B.C., "If water is poured on level ground it will avoid the parts which are dry and move towards those

6

that are wet. If [two] identical pieces of firewood are exposed to fire, the latter will avoid the damp and ignite the dry one. All things reject what is different [from themselves], and follow what is akin."[13]

But, distrusting reason as they did, the Taoists neither developed these theories nor welded them into a systematic account of nature comparable with that of Aristotle. As a result Chinese science remained intellectually fragmented, capable of steady empirical accumulation in many fields but incapable of further theoretical growth. For example, none of the theories we have mentioned stimulated a development like that of the "impetus" tradition in dynamics. Aristotle had maintained that a body in forced motion continues to move only so long as it is in contact with the original mover. If so, he was asked, why does an arrow continue to fly for some time after it has been released? The answer, he replied, is that the air displaced by the arrow as soon as it is fired rushes behind it and thrusts it forward. But, as John Philoponus of Alexandria objected in the sixth century A.D., there is no reason why the air should move behind the arrow rather than in some other direction. "How is it," he wrote,

> that the air, pushed by the arrow, does not move in the direction of the impressed impulse, but instead, turning about, as by some command, retraces its course? Furthermore, how is this air, in so turning about, not scattered into space, but instead impinges precisely on the notched end of the arrow and again pushes the arrow on and adheres to it? Such a view is completely implausible and is more like fiction.[14]

The arrow, he concluded, continues in flight because a force—later called "impetus"—is imparted to it by the archer and remains with it after it has left the bow. This theory of projectile motion was further developed by a succession of Islamic and medieval philosophers. Chinese science has no parallel.

Backward European science began its meteoric career with Galileo's discovery that mathematical hypotheses, tested by experiments, can give precise knowledge of the workings of nature. This approach, together with the philosophy of mechanism (the doctrine that all natural phenomena can be explained in terms of the motions of particles under the influence of forces), soon put European science

7

far ahead. For it turned out that, contrary to the Taoists, the order of nature is not after all inscrutable. What remains to be explained is why the Chinese themselves, in more than a millennium of inquiry, failed to hit upon the mathematical-experimental method and the mechanist philosophy.

A number of answers have been proposed. It has been suggested that the enormous Chinese bureaucracy with its Confucian ethos frustrated scientific innovation. The civil service examination was open to all and offered good careers to those who survived the intense competition. But it required a mastery only of the Confucian classics and literary works and so provided no incentive to study science and technology. Nevertheless, although the bureaucracy undoubtedly acted as a brake on theoretical inquiry, it did stimulate applied science and, in astronomy at least, systematic observation. It was responsible for the invention of the seismograph, the erection of rain- and snow-gauges, and the mounting of great expeditions to survey a meridian arc for 1,500 miles from Indochina to Mongolia and to map the stars of the southern hemisphere from Java. The Astronomical Bureau lasted 2,000 years without radical change. Its main functions were to record all celestial events and to forecast the fates of rulers and states from astrological omens. This arrangement insured a continuous flow of accurate data but discouraged original thinking and an interest in new problems.

A related suggestion is that the Chinese bureaucracy minimized the influence of the merchants. For over 2,000 years the civil service attracted the best brains in the country. Such was its prestige that even the sons of wealthy merchants struggled to get into it. Yet a flourishing mercantile class was essential, it has been argued, to the rise of modern science in Europe. The merchants had a financial interest in technological invention; they believed in the freedom necessary for scientific debate; and, being ready to work with their hands, they recognized the importance of experimentation. This argument is persuasive, but it should not be pressed too far. It has not been shown, for example, that scientific advance depends on whether or not a merchant class has won political power.[15] In Italy, for instance, republican systems of government, supported by the merchants, had widely given way to one-man rule before Galileo was born (1564). In

England, on the other hand, the Scientific Revolution began before the Civil War (1642–1648) and before the restoration of the monarchy (1688) by Parliament had consolidated the merchants' influence.

Another proposal is that the Chinese lacked the idea of a divine lawmaker and so never realized that nature has laws.[16] There is no doubt that this idea lent self-confidence to European science. If the universe is divinely designed, it is comprehensible and can be analyzed like a piece of machinery to see how it works. In revealing this design, science pays homage to its Creator. Nevertheless, as Needham points out,[17] the Chinese cosmos seems quite as rational as the European. It has the harmony of a pattern, not of a machine. This harmony is not imposed but arises from an inner necessity. All things in nature cooperate spontaneously because it is intrinsic to them, as parts of a pattern, to do so. For the Chinese, social and world order rested not on authority but on interdependence. In the words of a commentary on the *I Ching* (Book of Changes), "No one was ever seen to command the four seasons, yet they never swerve from their course."[18]

Again, Europeans inherited the union of Babylonian observational astronomy and Greek geometry in the research tradition of Ptolemy. They also received the rich mathematics of Islam, including algebra, which was essential for the invention of calculus and the creation of Newtonian mechanics. Mathematical astronomy, which spearheaded the Scientific Revolution, was a weak area in Chinese science. Yet it is surely implausible to claim that modern science could only have begun in astronomy. It might be argued equally well that Chinese science could have advanced through the study of magnetism and electricity to the field physics of Faraday and Einstein without undergoing a Newtonian phase. The laws of these phenomena could have been stated in traditional Chinese mathematics, and Chinese geometry could have been refined to handle astronomy later.

Far from being predestined, modern science seems to have arisen in Europe out of a fortuitous combination of historical conditions. The Renaissance, for instance, fostered individualism and an interest in this world rather than the next. The Reformation and

Counter-Reformation weakened the hold of established religion and reduced religious opposition to secular enterprises. Capitalism created a class with an appetite for new knowledge, a sympathy with experimentation, and a belief in the exploitation of nature. Voyages of discovery enlarged the known world and revealed a wealth of novel phenomena. The notion of a divine legislator made science self-confident and respectable, and the legacy of Ptolemaic astronomy and Arabic mathematics provided the conceptual tools for a breakthrough. A native tradition of experimentation began with the artisans and alchemists of the Middle Ages and was then broadened by the wars of the sixteenth century, which stimulated educated men to master the technologies of gunnery and fortification. The diversity of Europe, with its many peoples, tongues, and traditions, meant that a climate unfavorable to science in one country could be balanced by a congenial climate elsewhere.

But though these conditions were sufficient for the birth of modern science, they were not essential. The combination of mathematical hypotheses and experimental testing calls for a knowledge of mathematics and a tradition of experimentation, which is enough to explain why modern science was not born in New Guinea. But where these are present, as in China, a similar outcome could follow from many quite different sets of factors. Explanations of China's failure to create modern science do not prove that there is a single route to that science but only that China did not take the route we did. From now on, then, we shall take Western science as our paradigm, not because it is the only science, but because it has been the most successful.

RESEARCH TRADITIONS

Science aims at giving a complete explanation of nature's order. To realize this aim it proposes and tests theories that attempt to explain particular aspects of this order. A scientific theory is a set of statements describing the nature and/or operation of an unobserved entity or process postulated as the cause of certain observed facts. This entity or process is regarded as a possible hidden order, or "mechanism," whose existence may be verified by checking whether the facts occur as the theory predicts. When the theory has been well confirmed, the order it postulates is held to be real. Thus electric and magnetic fields, elementary particles, genes, and natural selection, whose existence once was bitterly contested, are now regarded as actually present in nature.

Most theories are proposed within a research tradition. Great theories in fact create such traditions. A research tradition is a succession of investigations undertaken by a number of scientists under a set of general assumptions. These assumptions state what the fundamental entities are in a domain and how they interact. According to the Platonic-Ptolemaic tradition, for instance, the planets are perfect spheres and move in circles about the earth at constant speed. According to the Newtonian tradition, the world is composed of infinitesimal particles interacting across empty space through forces of attraction and repulsion. A research tradition often will specify how the relevant phenomena are to be investigated and how theories should be constructed to explain them. Thus in the Ptolemaic tradition all celestial phenomena were supposed to be explained by as few circular motions as possible. In the Newtonian tradition theories came to be written in the form of differential equations.

A research tradition stimulates the creation of a string of theories. A theory explains the behavior of certain phenomena within the domain by postulating at least some of the assumptions of the tradition and by making further more specific assumptions of its own. Take the nineteenth-century kinetic tradition, an offshoot of the Newtonian tradition. Scientists in the kinetic tradition assumed that heat is caused by the random motions of molecules, which make up matter in all its forms. The implications of this thesis were worked out in the study of gases. It was assumed that a gas is a swarm of constantly moving particles governed by the laws of Newtonian mechanics. In the early years of the tradition, theories were put forward by August Krönig, Rudolf Clausius, James Clerk Maxwell, and Ludwig Boltzmann.[19]

In 1856 Krönig, a little-known Berlin physicist, proposed that the molecules of a gas are smooth elastic spheres traveling in straight lines at the same speed. From these simple assumptions he made a

11

number of deductions, including the Boyle-Charles gas law ("at a constant temperature the volume of a gas varies inversely with the pressure") and the proposition that gases are warmed by compression because the moving piston accelerates the molecules.

Krönig's theory was a challenge to Clausius, codiscoverer of the second law of thermodynamics ("the entropy of the world tends to a maximum"). More realistic than Krönig, he proposed that the elastic molecules rotate, vibrate, and collide, traveling in different directions at different speeds. Nevertheless, he argued, when a gas reaches a steady state, there is a mean speed and a mean free path (the distance a molecule travels without striking another), both of which he calculated.

Clausius's work stimulated the Scottish physicist Maxwell to produce a better theory based on statistics. In addition to the assumptions of the tradition, Maxwell made three assumptions of his own: first that the molecules are elastic spheres, second that after a collision all directions of rebound are equally likely, and third that each component of a molecule's velocity is independent of the others (the distribution function). From these assumptions he deduced the mean free path of a molecule, a law for the distribution of velocities among molecules, and the proposition that the kinetic energies of the molecules tend to become equal. He then made some important predictions about the properties of gases—for instance, that the viscosity (or internal friction) of a gas is independent of the density, and that it varies with the square root of the temperature.

Experiments refuted the second prediction, however, and indicated that viscosity varies with temperature as such. Maxwell then produced a second theory in which he assumed that the molecules are centers of force rather than elastic spheres. This assumption implied that the distance traveled by two molecules approaching a collision varies with their velocity and hence with the temperature. (If the temperature increases, so does the mean free path and the viscosity.) Maxwell also assumed that the velocities of two colliding molecules, rather than the velocity components of a single molecule, are statistically independent. He then deduced the distribution function which he had previously assumed, together with the main properties of gases in equilibrium, such as the fact that the volumes of two com-

12

bining gases bear a single numerical relation to each other (Joseph Gay-Lussac's law of 1808).

Struck by Maxwell's statistical techniques, the Austrian physicist Boltzmann, early in his career, undertook to extend some of the former's deductions. Abandoning Maxwell's assumption that gas molecules collide, he assumed simply that there is a fixed amount of energy to be distributed among a finite number of molecules in such a way that all combinations of energies are equally probable. He then made two important deductions: an expanded version of Maxwell's law for the distribution of velocities, covering polyatomic molecules under the influence of forces, and a general equation for all forms of molecular interaction.

These examples show how scientists work from the assumptions of a tradition to create new theories.

Much of the history of Western science consists in the formation, growth, and decline of competing and complementary research traditions. From Hellenistic times until the Renaissance for example, the Aristotelian tradition in physics and biology, the Ptolemaic tradition in astronomy, and the Galenic tradition in medicine largely complemented one another. In the eighteenth century, on the other hand, in the physical sciences three traditions competed for supremacy: the Cartesian, the Leibnizian, and the Newtonian.[20] Descartes maintained that the universe consists of solid particles interacting by contact. Leibniz proposed that the universe is a field of force, of which material bodies are intensive manifestations. However, the Newtonian tradition was by far the most successful, providing some of the assumptions for virtually all research in physics until the end of the nineteenth century.

Research traditions evolve in three main ways: by creating new theories, by changing their assumptions, and by uniting with other traditions. Through theory-creation a tradition can yield a staggering wealth of implications. Take the quantum tradition. This originated in the attempts of Albert Einstein and Max Planck to explain the emission and absorption of radiation by matter. In 1913 Niels Bohr used Planck's notion of quanta (chapter 7) to explain the spectral lines (chapter 4) of the hydrogen atom. By 1926 the Bohr tradition, which mingled Newtonian and quantum concepts, had been re-

13

placed by the mature quantum tradition based on the mathematical equivalence of Erwin Schrödinger's wave mechanics and Werner Heisenberg's matrix mechanics. Heisenberg also pointed out that the position and momentum of a subatomic particle cannot be measured simultaneously with complete accuracy. The new tradition was able to explain all the phenomena that had perplexed Newtonian physicists at the beginning of the century: black-body radiation (chapter 7), Brownian movement (chapter 1), the photoelectric effect, X-ray diffraction, the Compton effect, and so on.[21]

The quantum tradition then promoted research in two directions. On the one hand, it stimulated the solution of more and more complicated problems about solids, liquids, and gases, treated as structures of electrically charged particles. On the other, it sought to discover the ultimate constituents of matter in the form of ever more elementary particles. Let us look at the second development.

In 1928 Paul Dirac united quantum mechanics with special relativity in quantum electrodynamics, the first of the quantum field theories, in which particle interactions are explained by the exchange of particles themselves. The addition of special relativity enabled scientists to predict the behavior of particles of high energies and velocities. In the early 1930s the positron, neutron, and neutrino were discovered, and in 1935 Hideki Yukawa predicted the existence of the meson. During World War II, particle accelerators and photographic techniques were improved. As a result, by the late 1950s about thirty particles were known. Physicists still hoped that their interactions could be explained by a revised version of quantum field theory. Shortly before 1960, however, the big accelerators went into action, and within a few years more than 200 particles had been found. (For a while they were turning up by the week!) Attempts were made to classify them, notably by Murray Gell-Mann and Yuval Ne'eman. Theories proliferated, but fell into two main classes, "fundamentalist" and "bootstrap." The first kind, variants of the quantum field approach, were the most common. They assert that all particles are made out of one or a few basic particles, such as Gell-Mann's quarks. Bootstrap theories, on the other hand, claim that all particles are equally fundamental and all are made out of one another— "nuclear democracy." The most successful of these is the S (scattering)-matrix theory, proposed in the 1950s by Geoffrey Chew, following an earlier suggestion by Heisenberg.

What an extraordinary development! Who could have predicted in 1917 that within 10 years quantum theory would have explained matters as diverse as the structure of a piece of metal, the behavior of electrons, and the extent to which the world could be measured at all? Still less, who could have forecast its fundamental conceptual breakthrough, the mathematical expression of the revolutionary idea that the microscopic constituents of the world have a dual wave-particle nature? Yet again, in 1926 who could have foreseen that 30 years later quantum physicists would be seeking a theory to explain the four forces of the universe (gravitational, weak, electromagnetic, strong) through which all particles interact?[22]

A research tradition also may evolve by changing its assumptions, as did the Mendelian tradition in genetics. This tradition was founded in 1903–04 by the American Walter Sutton and the German Theodor Boveri, who proposed that the patterns of inheritance described by Mendel result from the transmission of units of heredity ("genes") located in or on the chromosomes. These patterns are to be investigated, they said, through artificial breeding. In the 1910s and 1920s Thomas Hunt Morgan and his associates, using the Drosophila fly, showed that the genes are indeed "in" the chromosomes. Then in the 1950s, thanks to the development of biochemistry and the invention of more sophisticated genetic techniques, it became possible to study inheritance patterns in microorganisms such as bacteriophages (viruses that infect bacteria). Classical genetics now became transmission genetics. The assumptions of the tradition were reformulated so as to apply to molecular phenomena, such as loss of enzyme activity. The notion of a gene also was redefined. The classical gene occupied a definite spot on the chromosome. In transmission genetics, on the other hand, the gene includes all those units, wherever located, that act as parts of a single gene. Transmission genetics also introduced new types of gene, such as the regulator and the repressor. Finally, it contradicted Mendel's law of segregation, that in the heterozygote the alleles within the same gene do not mix. In transmission genetics, units as small as nucleotides act as alleles and may recombine within a single gene.[23]

15

As an example of development through unification, take the union of Darwinian evolution with Mendelian genetics in the synthetic theory of evolution. Darwin argued his case from facts well known at the time: the tendency of organisms to overproduce, the tendency of populations of organisms to remain constant, and the tendency of individuals to vary. He pointed out that since all species tend to overproduce, and yet their numbers remain constant, there must be a struggle for existence. So, he concluded, since species exhibit individual variation, some variants must survive and perpetuate themselves while others are eliminated. In short, the fittest survive.

Darwin's theory swept paleontology, anatomy, and embryology, but was criticized by other biologists for its failure to explain the source of individual variation. However, when Darwin published *The Origin of Species*, Gregor Mendel was experimenting with pea plants in the Augustinian monastery at Brünn, Austria. The laws of heredity he derived from these experiments were ignored until 1900, when they were independently rediscovered by three biologists. However, the early Mendelians regarded genetic changes in individuals as the main cause of evolution and relegated natural selection to the minor role of eliminating those forms that failed to change. It was not until the 1930s that evolutionists saw in population genetics a viable explanation of evolutionary change. Ronald A. Fisher and J. B. S. Haldane in England and Sewall Wright in the United States worked out a mathematical theory of population genetics related to evolution. The modern theory of evolution thus is a synthesis of natural selection and population genetics, which asserts that evolution consists in changes in the gene pool of a population. The gene pool (the totality of genes in a population) is changed mainly by mutation. Where, for Darwin, natural selection had been differential survival, for the synthetic theory it is differential reproduction. Natural selection implies that genes which increase the reproductive success of the individuals carrying them will be transmitted to the following generation more frequently than their alternates. In a relatively few generations these genes will spread throughout the population. Natural selection has been compared to a sieve which retains the rarely occurring useful mutant and lets go the more common harmful ones. But natural selection does more than this. It also acts creatively by multiplying adaptive combinations of genes which would not have multiplied otherwise.

HISTORICALITY AND THE SCIENTIST

Science is historical in the sense that it is an activity, an institution, and a body of knowledge that change in time as a function of the quest for a complete explanation of nature's order. I have considered this process in terms of the evolution of research traditions. Let us now look at it from the point of view of the individual scientist.

The scientist may refine existing knowledge or he may produce fundamentally new knowledge. He refines knowledge by making it more precise and more certain. One way of attaining precision is to measure the properties of phenomena more accurately, especially if a new experimental technique makes possible a more stringent test. For example, the proportionality of inertial and gravitational mass[24] was measured in successively more precise experiments by Isaac Newton (1686), Friedrich Bessel (1823), Roland von Eötvös (1922), and Robert Dicke (1964)—by Dicke to an accuracy of 10^{-11}.[25] Another way is to clarify and reformulate concepts and statements, often by expressing them mathematically. During the eighteenth and the early nineteenth century, for instance, several generations of Europe's leading mathematicians—Daniel, Jakob, and Johann Bernoulli, Jean d'Alembert, Leonhard Euler, Joseph Lagrange, William Rowan Hamilton—rewrote and extended Newton's mechanics, making it simpler and clearer.

The scientist seeks certainty by accepting only those hypotheses that have been tested as rigorously as possible.[26] This attitude is exemplified in the response of scientists to Einstein's general theory of relativity (proposed in 1915). For nearly 50 years general relativity was largely ignored on the grounds that it was too complicated mathematically, could not be tested in the laboratory, and was incompatible with quantum mechanics. Only a few researchers, devoting their careers to the theory, elucidated its extraordinary consequences, most of which Einstein never knew. One of these consequences is that many huge stars are destined to collapse under the weight of their

17

own mass and disappear, leaving behind "black holes" from which nothing, not even light, can escape. Another is that the universe contains singularities, places where space-time begins or ends and the known laws of physics break down, as inside black holes and at the birth of the universe.

Then the 1960s brought a series of spectacular discoveries in space: exploding galaxies, quasars, cosmic microwave radiation from the Big Bang, pulsars, and compact X-ray sources. (Quasars are starlike objects that must outshine entire galaxies if they are as distant as the red shift of their spectra suggests; pulsars are the rapidly blinking debris of supernova explosions, thought to be ultradense neutron stars; compact X-ray sources may be neutron stars or else black holes denser still.) At the same time a host of new techniques was invented for testing gravitational theories: radar trained on planets and satellites, laser beams aimed at the moon, atomic and molecular clocks, gravimeters, long-baseline interferometers, and many others.[27] As a result of these developments—new observations, new instruments— gravitation theory and astrophysics currently are the most exciting fields in physics. There is a boom in general relativity. Theorists are deriving further theorems from it, and experimenters are devising test after test to decide between the theory and its many competitors.[28]

The scientist produces fundamentally new knowledge by proposing theories that are broader, deeper, and simpler than their predecessors. A theory is broader when it explains a greater range of phenomena than its rival. Breadth sometimes is achieved by combining two existing theories or sets of laws within a more comprehensive theory, as Newton combined the laws of Kepler and Galileo, among others, and Maxwell integrated the sciences of optics and electromagnetism. Or it may be achieved by explaining facts in a wide spread of fields, as Darwin explained data throughout the life and earth sciences.

A theory is deeper when it proposes a mechanism to explain the mechanism postulated by the theory it replaces. Contrast Einstein's theory of gravitation with that of Newton. The latter states that gravitation is an instantaneous attraction between bodies in absolute space unaffected by matter. Einstein explains this mechanism by proposing that space is warped by matter, and that the resulting curvature deflects bodies from their paths in the direction of bodies more massive than themselves. Or contrast Darwinian evolution with the synthetic theory. Darwin explains the evolution of species as being due to the action of natural selection on spontaneous variation. But the synthetic theory goes deeper, because it explains the variation itself as an outcome of gene and chromosome mutation. I cite Schiller's couplet, of which Niels Bohr was especially fond:

> Nur die Fülle führt zu Klarheit.
> Und im Abgrund wohnt die Wahrheit.

"Only fullness leads to clarity, and truth dwells in the depths."[29]

Finally, a theory is simpler than a rival when it has fewer premises relative to the number of consequences deduced from them. The theory of special relativity is especially simple and daring, being based on two general principles, light and relativity. The first asserts that the speed of light is constant, the second that the laws of nature are the same in all reference frames moving uniformly relative to one another.[30]

The Advance through Problems. Science also is inherently historical in that it tends to be cumulative. Every investigation is an attempt to solve a problem arising from the solution of a previous problem. If successful, it opens up one or more new problems for research to investigate. The solved problem is a link in the chain of problems and their solutions through which science advances. A new theory generally is a fertile source of problems through the predictions it yields.

Problems may be empirical or theoretical.[31] An excellent example of the first is the problem of Brownian motion. The random motion of tiny particles suspended in a fluid in thermal equilibrium was first discussed by the Scottish botanist Robert Brown in 1827. For the rest of the century scientists debated whether it was a serious or even a genuine problem, and who ought to solve it. During the 1830s and 1840s, for instance, it was alternately regarded as a biological problem (the particles being perhaps "animalcules"), a chemical

19

problem, a problem in optical polarization, in electrical conductivity, in heat theory, an uninteresting mechanical problem, and a non-problem. Toward the end of the century it was recognized as a serious anomaly for the laws of thermodynamics and the thermodynamic research tradition. Finally it was solved by Einstein and Jean Perrin in a striking comeback by the kinetic tradition.[32]

Theoretical problems may be internal or external to a theory. Quantum mechanics, for instance, despite a multitude of successful predictions, is plagued by foundational problems. One of these is the puzzling fact that under some conditions, such as interference and diffraction,[33] radiation behaves like a wave process governed by Maxwell's equations of the electromagnetic field, and under others, such as the photoelectric effect and Compton scattering, it behaves like a beam of particles, quanta of energy called photons. The theory can predict the outcome under both sets of conditions but cannot explain it. The bewilderment of physicists at this mysterious dualism was well expressed by Sir William Bragg in his famous remark, "We use the classical theory on Mondays, Wednesdays, and Fridays and the quantum theory on Tuesdays, Thursdays, and Saturdays."

An example of an external problem is the conflict between the Ptolemaic theory and the Platonic principle of celestial motion which the theory claimed to respect. Like quantum mechanics, the Ptolemaic theory was enormously successful empirically. But to achieve this success it had to violate the principle of perfect circular motion by assuming, for instance, that certain planets move round empty points in space, that planets do not always move at constant speed, and so forth. This conceptual disharmony was the chief fault that Copernicus found in the theory.

The scientist must tackle a problem with the data and techniques at his disposal, and they may well be inadequate. Nevertheless, partial solutions lead to better ones. Bohr's theory of the hydrogen atom was an inconsistent compromise between classical and quantum physics,[34] but it stimulated hypotheses and uncovered facts that led to the mature quantum theory of the mid-1920s. By the same token, a scientist who formulates and solves an important problem which is not recognized at the time may well find his work ignored or underappreciated. Daniel Bernoulli's explanation of gas pressure anticipated the kinetic theory of gases by a century. William Prout's hypothesis, that the atomic weights of the elements are multiples of the weight of the hydrogen atom, was resisted by most chemists until its vindication a century later by Ernest Rutherford and Frederick Soddy using new experimental techniques. Mendel's paper on the laws of genetics lay buried in an obscure journal for 34 years before it was rediscovered. Alfred Wegener's hypothesis of continental drift was vehemently rejected for half a century until favorable evidence emerged from the new sciences of rock magnetism and ocean ography.

OTHER ASPECTS OF HISTORICALITY

History and Corrigibility. Science is historical, again, in that every scientific statement or set of statements is open to revision or replacement in the light of further evidence or new ideas. Some theories—for example, Maxwell's electromagnetic theory—have established themselves over an extended period of testing and refinement, and most scientists would fight hard to retain them. Nevertheless, even these will be abandoned if they are decisively refuted, and if alternative theories are available that promise to explain the facts more satisfactorily. Because all scientific conclusions ultimately are tentative, science always can criticize and transform itself. If Newtonian mechanics had not been regarded as ultimately replaceable, relativity theory and quantum mechanics would not have been invented to replace it. (Newton's theory still is held to account correctly for the motions of macroscopic bodies at speeds considerably less than that of light.) Because science is tentative it can be revolutionary.

Objectivity. As science has grown, it has become less anthropomorphic and more objective. This is to be expected, as Nicholas Maxwell points out,[35] for if the universe is intelligible, it is so potentially to all people and not merely to those having certain sense

21

organs and coming from certain cultures. The more simply and beautifully our theories explain the natural order, the wider the range of human beings to whom they should be intelligible. This trend toward objectivity may be seen in Copernicus's denial that the earth is the center of the universe, and in the seventeenth-century demand for explanations in terms of primary qualities (secondary qualities being important only to human beings with their special sense organs and nervous system). It may be seen, too, in Galileo's principle that uniform, unaccelerated motion occurs relative to the observer, a principle that denies the more parochial Aristotelian idea that the earth is absolutely at rest and hence is a privileged frame of reference from which all motions anywhere may be accurately observed. It is visible again in Einstein's broader principle of relativity. This principle maintains against Newton that there is no special set of reference frames for formulating the laws of nature,[36] and that the laws therefore must be given the same form in all reference frames, that is, for all observers in arbitrary positions and relative motion.

Explanation. What, then, are we to conclude about the world from our study of the advance of science? The more we understand the world, the more strange and also the more simple it appears. In explaining the world we do not make it more familiar, for the familiar is the anthropomorphic and the socially and culturally specific. As Niels Bohr so keenly observed,[37] "If a man does not feel dizzy when he first learns about the quantum of action, he has not understood a word." We therefore should reject the idea that the more science grows, the less it explains, its theories improving only as instruments for predicting phenomena. On the contrary, the more bizarre and the more simple science reveals the world to be, the better it explains the world. What, for instance, could be more bizarre and yet more simple than a black hole—a gigantic star that has collapsed so far that its enormous gravity will not allow even light to escape from it, with the result that the star is invisible? Physicist John Archibald Wheeler has made this point well. "The universe," he says, "is far stranger and more beautiful than we realize, and also far more simple. But we have no hope of seeing how simple it is until we first recognize how strange it is."[38]

SUMMARY

Science is intrinsically historical. Owing to the limitations of the human mind, the scientific mission—to give a complete explanation of the natural order—will take many centuries to fulfill and indeed may never be fulfilled. In all civilizations some men have sought to explain this order naturalistically. Nevertheless, until modern times they did not regard themselves as scientists or as contributors to a supracultural tradition. Natural knowledge was diffused by accident rather than by design. Hence we see in the past several scientific endeavors, each evolving in a different civilization, rather than a single historical movement in which every civilization took part. Over a millennium the Chinese in particular built up a large body of empirical knowledge. If circumstances had been different, they might have arrived at a modern, field theoretic approach to nature. Instead, modern science—mathematical, experimental—took off in seventeenth-century Europe by way of mechanics.

Science grows largely through the evolution of research traditions. These guide research by stating the nature and interactions of the fundamental entities in a domain. They develop in three main ways: by creating new theories, by changing their assumptions, by uniting with other traditions. Within traditions, and sometimes outside them, scientists refine knowledge, making it more precise and certain. They also produce fundamentally new knowledge by proposing theories that are broader, deeper, and sometimes simpler than their predecessors.

At the level of the individual and of the research tradition alike, science advances by solving problems, empirical or theoretical. Since its solutions are tentative, they always can be reconsidered and replaced by new ones. Thus science is committed to self-criticism and the possibility of further growth. As science grows, it tends to become less anthropomorphic and more objective, and to disclose an

23

order in nature that is both stranger and simpler than we had supposed.

So far I have assumed that over centuries science has advanced successively closer to the truth of nature. But this assumption has been challenged, and in the next chapter I shall consider whether we any longer are entitled to make it. I also shall inquire whether science is likely to continue growing indefinitely.

NOTES

1. The Lyceum was a grove just outside Athens where teachers gave lectures. Aristotle's school seems to have acquired property there shortly after his death in 322 B.C. The name seems to have been given to the school itself because it was the largest research community in the ancient world until surpassed in the next century by the scientific community of Alexandria.

2. See Joseph Needham, *Science and Civilisation in China:* vol. 2, *History of Scientific Thought*, pp. 63–68.

3. Quoted by G. E. R. Lloyd, *Greek Science After Aristotle*, p. 21.

4. *Ideas and Opinions* (London: Alvin Redman, 1954), pp. 39–40.

5. George Meredith, a nineteenth-century man of letters, was the author of some vivid, idiosyncratic novels of manners, such as *The Egoist* and *The Ordeal of Richard Feverel*.

6. See Joseph Needham, *The Grand Titration: Science and Society in East and West* (Toronto: University of Toronto Press, 1969), pp. 15–16, and, with the collaboration of Wang Ling, *Science and Civilisation in China;* vol. 3, *Mathematics and the Sciences of the Heavens and the Earth*, pp. 447–51.

7. *The Grand Titration*, pp. 52, 58, 76.

8. In the sixteenth century Pierre de Fermat, a member of the provincial assembly of Toulouse, proposed that the actions of nature always take the least possible time.

9. I choose China because science was practiced longer and more successfully there than in any other ancient civilization. Nevertheless, the same questions could be asked of other civilizations.

10. In *Science and Civilisation in China*, 4 vols. to date.

11. A *nova* is a star that increases its light and energy up to a million times in a few days and then after several months or years returns to its former obscurity. A large nova may emit as much energy as the sun does in 10,000 years. Indeed, if the sun became a nova, the earth would be destroyed in a few hours or days. Some 30 novae have been observed in the past few hundred years.

12. Needham, *The Grand Titration*, pp. 46 50.

13. Quoted by Needham, *Science and Civilisation in China*: vol. 2, *History of Scientific Thought*, p. 281.

14. Quoted by Lloyd, *Greek Science after Aristotle*, pp. 158–59.

15. A. C. Graham, "China, Europe, and the Origins of Modern Science," in *Chinese Science: Explorations of an Ancient Tradition*, ed. Shigeru Nakayama and Nathan Sivin (Cambridge, Mass.: MIT Press, 1973), pp. 54–55.

16. Needham, *The Grand Titration*, ch. 8.

17. *Science and Civilisation in China*: vol. 2, *History of Scientific Thought*, pp. 287, 290, 562, 582; see also Graham, "China, Europe, and the Origins of Modern Science," pp. 32–34, 58.

18. Quoted by Needham, *Science and Civilisation in China*: vol. 2, *History of Scientific Thought*, pp. 287, 561.

19. See Peter Clark, "Atomism versus Thermodynamics," in *Method and Appraisal in the Physical Sciences*, ed. Colin Howson, pp. 45–63.

20. See Yehuda Elkana, "Newtonianism in the Eighteenth Century" (essay review), *British Journal for the Philosophy of Science* 22 (August 1971): 297–306.

21. The *photoelectric effect*—the emission of electrons by metal surfaces when struck by light—was explained by Einstein in 1905 on the assumption that light is particulate. The incoming photons of light, he said, dislodge the electrons on impact. *X-ray diffraction*: When X-rays are directed at a crystal, they alternately reinforce and cancel one another (like waves meeting an obstacle), producing parallel light and dark bands. In 1912 in a celebrated experiment the German physicist Max von Laue trained a beam of X-rays on a crystal of zinc sulfide. He showed that X-rays are waves and that the atoms of a crystal are arranged regularly in lattices, which are responsible for the interference pattern. *Compton effect*: When X-rays and other strong electromagnetic radiations are scattered by electrons, they increase in wavelength. In 1922 the American physicist Arthur Holly Compton proposed that this occurs because X-rays consist of photons. When the incoming photons collide with the electrons, new photons of lower energy and longer wavelength are produced,

which scatter at angles varying with the amount of energy lost by the recoiling electrons.

22. Gravity is the weakest of these forces. The weak force itself, which is observed in the radioactive decay of atomic nuclei, is a billion times less intense than the electromagnetic force, which binds electrons to nuclei in atoms and molecules. The strong force, which binds protons and neutrons in the nucleus, is more than 100 times stronger than the electromagnetic force and acts only within distances of a fermi (a millionth of a millimeter). On the development of quantum field theory, see Steven Weinberg, "The Search for Unity: Notes for a History of Quantum Field Theory," *Daedalus* 196 (Fall 1977): 17–36.

23. An allele is either one of a pair of alternative Mendelian characters, such as tallness and shortness in a pea plant. A heterozygote is an animal or plant with two different alleles for a specific trait, e.g., one for tallness and one for shortness.

24. Inertial mass is the measure of a body's resistance to being accelerated. Gravitational mass is the measure of the force of attraction exerted by or on the body.

25. Physicists often use powers of 10 as a convenient shorthand for very large or very small numbers; 1,000, for instance, becomes 10^3 and one millionth becomes 10^{-6}.

26. See Gary Gutting, "A Defense of the Logic of Discovery," *Philosophic Forum* 4 (Spring 1973): 393–94. Certainty and precision cannot be attained absolutely, however, but only increased.

27. The *red shift* is the displacement of light from stars and other celestial objects toward the red end of the spectrum. The displacement, which corresponds to an increase in wavelength, is proportional to the speed at which the objects are receding from the earth. The red shift is the most striking evidence that the universe is expanding. Much brighter than a nova, a *supernova* breaks apart and does not return to its previous state. It is thought to turn into a neutron star (a collapsed star of terrific density), so-called because under extreme pressure and temperature its electrons and protons fuse into neutrons. A *gravimeter* is an instrument for measuring variations in the earth's gravitational field. A *long-baseline interferometer* is a radio telescope with separate antennas—receiving radio waves from the same source—placed thousands of miles apart and connected to a single receiver.

28. See Clifford M. Will, "Gravitation Theory," *Scientific American* 231 (November 1974): 25–33.

29. From the Sonnets to Confucius. Quoted by David Park, "Time and Form in the Physical World," *Boston University Journal* 24 (1976): 32.

30. A reference frame also is known as a coordinate system. The position of an object in space usually is plotted with the aid of three coordinates, which are measured from the location of an observer. Einstein added a fourth coordinate—time—and maintained that space and time are relative to the observer. For Einstein, then, all

coordinate systems are equally valid, and the laws of nature must take the same form in all of them.

31. See Larry Laudan, *Progress and Its Problems*, chs. 1, 2.

32. *Ibid.*, pp. 19–20.

33. *Interference* is a wave phenomenon; when two light beams from the same source meet, their waves variously neutralize and support one another, producing alternate light and dark bands or fringes. *Diffraction:* an interference pattern which occurs when light passes by the edge of an opaque body or through a narrow slit or is reflected from a ruled surface.

34. Bohr explained the intraorbital motion of the electron around the atomic nucleus by means of Maxwell's equations, which apply to continuous motion. He explained the electron's movement from one orbit to another (its "jumps") by means of Planck's quantum theory, which applies to discontinuous motion.

35. "The Rationality of Scientific Discovery, Part II: An Aim-Oriented Theory of Scientific Discovery," pp. 270–71.

36. According to Newton's first law, a body free of external forces should experience no acceleration. Therefore, there must be an observer for whom this statement is true. Hence there is a set of reference frames ("inertial frames") from which the motion of a forcefree body will be observed to be unaccelerated.

37. Quoted by C. F. von Weizsäcker, "The Copenhagen Interpretation," in *Quantum Theory and Beyond: Essays and Discussions Arising from a Colloquium*, ed. Ted Bastin (Cambridge: Cambridge University Press, 1971), p. 26.

38. Quoted by Maxwell, "The Rationality of Scientific Discovery, Part II," p. 271.

Chapter 2

Progress in Science

DOES SCIENCE PROGRESS?

Of course it does, most scientists would reply. Ever since the eighteenth century, science generally has been regarded as the paradigm of a progressive enterprise. Science is said to progress because it uses an increasing number of ever more precise techniques of investigation to accumulate an expanding stock of well-confirmed facts. The facts are summarized in a growing body of ever more accurate and comprehensive laws. These laws, in the physical sciences at any rate, are explained by means of theories of successively greater scope, simplicity, and precision, each theory entailing its predecessor as a logical consequence or special case. For example, geometrical optics [1] was explained by Young and Fresnel's wave optics, which, together with the laws of magnetism and electricity, was explained by Clerk Maxwell's electromagnetic theory, which in turn was explained by quantum mechanics.

Science, then, in the traditional view, is progressive in that it explains more and more facts and so captures more and more of the truth about nature. It also is cumulative, in that it adds to its past findings instead of replacing them.

But this view is too optimistic. As I shall show, science seeks to be progressive, and in the long run has been so, only in the sense that it endeavors to explain, and so far seems to have explained, more and more of the order of nature. It has not been continuously progressive, however, and it is only partly cumulative.

Why has science not been continuously progressive? Incorrect hypotheses sometimes have been preferred to correct ones. The idea that the earth moves around the sun was put forward by Aristarchus of Samos as early as the third century B.C., but was not advanced

28

again until 1543 A.D. by Copernicus. In the nineteenth century, as we have seen, Mendel's hypothesis was ignored and Prout's was rejected. In our own century Wegener's hypothesis was dismissed for 50 years. Individual sciences also have stagnated. After 200 A.D. little progress was made in Greek science outside medicine and astronomy. In certain countries sciences actually have regressed, as Russian genetics did under Lysenko. Again, science appealed little to the Romans, and in Western Europe during the Dark Ages it was abandoned entirely. Granted, modern science has been growing exponentially for more than two centuries, but, as we shall see, this trend cannot continue indefinitely. The most we can say is that science has been progressive so far, but only in the long term and despite many setbacks.

Nor is science always cumulative. Although many facts and laws, and the substance of certain theories, have been retained and refined, others have not. Some facts have been reclassified, some even rejected, when the theories that explained them were replaced by others that categorized the world differently. For example, when the affinity theory of chemical elements[2] was replaced by Dalton's atomic theory, many records of reactions were rewritten and some were thrown out altogether. Again, initially at least, some theories ignored certain problems solved by their predecessors.[3] Newtonian mechanics could not explain why the planets move in the same direction around the sun, a problem which had been solved by the Cartesian theory. Franklin's theory of electricity failed to account for the mutual repulsion of negatively charged bodies, a phenomenon explained by preceding theories.[4] Yet again, some theories, such as those of caloric, phlogiston,[5] and the inheritance of acquired characteristics, were rejected entirely, while others were transmitted only in part. For example, the substance of Newton's laws of motion is entailed by Einstein's special theory of relativity, but the Newtonian assumption that space is absolute and Euclidean is explicitly contradicted.

Nevertheless, science can be progressive even while rejecting some of its own achievements. Science can become more true, or at any rate less false, by refuting false theories as well as by retaining true ones. Aristotle's theory of a fixed scale of species of increasing perfection was refuted by Darwin's theory of evolution. Yet for more than 2,000 years it inspired a vast amount of fact-finding and much classification without which Darwin could not have conceived his theory.

Now let us look at the question of scientific progress from some other points of view.

VERISIMILITUDE

Karl Popper, a contemporary philosopher of science, argues that science is progressive, but on very different grounds from these. He holds that we never can know for certain whether a scientific theory is true or even probable because the theory always can be refuted by the next test. Nevertheless, it still is possible for a series of theories, even when refuted, to approach the truth successively. One theory may be less untrue than another, and we can guess fairly accurately which theory it is. Although both Kepler's and Newton's theories have been refuted, we can reasonably hold that Newton's is closer to the truth than Kepler's, and that Einstein's is closer than Newton's.

How does Popper justify this thesis? He maintains that the logical consequences of a theory can be divided into those that are true (the "truth-content" of the theory) and those that are false (the "falsity-content"). The difference between the two sets of consequences tells us the "verisimilitude" of the theory, or how close it is to the truth.[6] Given two rival theories A and B, we can guess that A has the greater verisimilitude if it entails all the true consequences of B and more beside, and if it has no more false consequences than B. Thus Newton's theory has more verisimilitude than Kepler's, and Einstein's than Newton's. Einstein's theory predicts all the facts predicted by Newton's, and predicts some (such as the motion of the planet Mercury) more accurately. In addition, Einstein's theory successfully predicts facts unanticipated by Newton, such as the shift of light, when emitted in a strong gravitational field, toward the red end of the spectrum. Since the logical consequences of a theory are infinitely many, we cannot in practice verify them all. Therefore, says Popper, the relative verisimilitude of competing theories must remain a guess. Even so, we can make the guess a good one by testing the theories as rigorously as we know how.[7]

MEANING-VARIANCE

Popper's verisimilitude thesis has been criticized on internal grounds,[8] but a more radical criticism, and one more relevant to our discussion, is contained in the so-called "meaning-variance" thesis proposed by Paul K. Feyerabend and Thomas S. Kuhn. According to this thesis, the meanings of scientific terms are "theory-laden" or determined by the theories in which they occur. How does this happen? For Kuhn and Feyerabend a term's meaning is its "reference," that is, the set of things to which the term is applied in virtue of their possessing the property associated with the term. The reference of the term "mass," for instance, is the set of things having mass. Now, at least some of the terms common to rival large-scale theories are used by each theory to make statements about objects or properties not mentioned by the other theory. It follows, on this view, that in each theory the terms have different meanings and consequently that the statements in which they appear cannot be compared. Thus Kuhn maintains that "in the transition from one theory to the next words change their meanings or conditions of applicability in subtle ways. Though most of the same signs are used before and after a revolution—e.g., force, mass, element, compound, cell—the ways in which some of them attach to nature has somehow changed. Successive theories are thus . . . incommensurable."[9]

For example, the terms "mass," "length," and "velocity" take on different meanings in classical and in relativistic mechanics. In classical mechanics these terms denote properties that are independent of the reference frame of the observer; in relativistic mechanics they denote properties that depend on it. In classical mechanics a body retains the same mass and length no matter how fast it travels relative to an observer. But in relativistic mechanics? Suppose, said Einstein, in a famous "thought experiment," you propel a clock that weighs one pound and is a yard wide. To an outside observer, the

31

clock, traveling at one-tenth the speed of light, 18,000 miles per second, will weigh 1.08 pounds, be two-tenths of an inch less than a yard wide, and run three-tenths of a minute slow. At half the speed of light it will weigh 2.25 pounds, be only 5 inches wide, and run 52 minutes slow. At .9999 of the speed of light it will weigh nearly 72 pounds, be half an inch wide, and nearly cease to tell the time. Thus, for Kuhn and Feyerabend, classical and relativistic predictions about the mass or length of an object are predictions about different properties.[10] Again, because the meanings of shared terms have changed, classical mechanics cannot be represented as a logical consequence or limiting case of relativistic mechanics. Although the statements in which Newton's laws normally are expressed can be derived under appropriate assumptions from the special theory of relativity, they have changed their meaning and therefore, says Kuhn, no longer express Newton's laws.[11]

What does this thesis imply for Popper's notion of verisimilitude? It says quite simply that since theories cannot be compared, we cannot know whether one has any more verisimilitude than another. To be sure, both Kuhn and Feyerabend allow for some intertranslation of theories. Some statements of one theory can be translated into the language of the other, or some statements of both theories can be translated into a third language. But even with partial translation, some of the known truth- and falsity-content of the theories cannot be compared. Hence we cannot make a good guess at the relative verisimilitude of the theories, and Popper's thesis appears to be refuted.

If we describe the world through the theories we hold about it, and if those theories are incommensurable, how do we know that science advances toward the truth? We do not, says Kuhn. Does science, then, make any progress at all? Yes, but only by increasing the number and precision of its problem-solutions, not by representing any more accurately what nature is really like. Newton's mechanics solves more problems than Aristotle's, and Einstein's solves more than Newton's, but neither Newton nor Einstein got any closer to the truth than Aristotle. The growth of science, says Kuhn, is like biological evolution. From primitive beginnings it moves onward, creating ever more specialized products, but it has no goal.[12]

MEANING AS INTENSIONAL REFERENCE

In the face of this criticism, how can a more traditional view of scientific progress be defended? Most obviously by proposing another theory of meaning. Mary Hesse maintains that if the meaning of a term such as "mass" depends on all the sentences in the theory in which it appears, then, trivially, it changes meaning from one theory to the next. She proposes, therefore, to define the meaning of a term as the "intensional reference" of that term, that is, as the class of things to which the term is applied insofar as those things are *observed* to be similar (as a result, perhaps, of measuring them in similar ways). On this interpretation, both Newton's and Einstein's theories apply the term "mass" alike to a range of phenomena. Einstein's theory asserts, for instance, and Newton's denies, that mass increases with velocity relative to the observer. Here the term "mass" is applied to the same phenomena; hence the theories are comparable, and Einstein's can be confirmed and Newton's refuted. On the other hand, Einstein ascribes "mass" to light, whereas Newton does not; and Newton ascribes "simultaneous" to certain pairs of events from which Einstein withholds it. Statements in the two theories, in which the same terms are applied to different phenomena, are logically incommensurable. Thus (contrary to Kuhn and Feyerabend) where two theories apply the same term to objects that are recognizably the same, they can be compared; otherwise not.[13]

How, then, does science progress? Chiefly, says Hesse, by accumulating laws, which can be translated from one theory to another insofar as they apply to the same classes of things.[14] Subject to slight corrections, a great many laws have remained stable through changes in theory: Archimedes' law of the lever (unrevised for more than 2,000 years), Snell's laws of light reflection and refraction, Newton's laws of motion, Boyle's law, Coulomb's laws of magnetic and electric interaction, Ohm's, Ampère's, and Faraday's laws of electricity, Mendel's laws of genetics, and so on. With the aid of these laws a multitude of experimentally controlled events have been made to happen and numerous technological artifacts have been created. These artifacts and events are striking evidence that the laws are true.[15] Nevertheless the artifacts, at least, do not validate the theories

33

by which the laws are explained. Science, insists Hesse, produces only phenomenal knowledge about observables—knowledge that can be applied technologically—and not theoretical knowledge about the hidden natures of things. Unlike laws, theories contradict and replace one another. Ever since the Greeks, for instance, science has oscillated between field and particle theories of matter. Hence theories remain hypotheses only and are not cumulative.[16] Technological devices are theory-independent because the laws by which they are built remain true regardless of how they are explained. Although nuclear reactors and transistors might be outmoded by a change in our theories of nuclear fission and the solid state, they would not work any the less efficiently. For the laws on which they rest would not have changed, only the explanation of them.

Here, however, Hesse confuses two modes of progress toward the truth and hence two forms of scientific knowledge. Laws represent the truth. Hence, as they accumulate, they represent more of it. Theories may be false, but are superseded by others that are less false. Hence, by replacement, theories come closer to representing the truth. For example, Roger Boscovich's theory (1763) that the ultimate constituents of matter are point centers of force is closer to the modern conception of the elementary particle than is the Newtonian notion of a solid, material atom, "a nice hard fellow," as Rutherford jokingly put it, "red or grey in color, according to taste."[17] Again, although theories of matter and light have altered between field and particle conceptions, both conceptions are combined (albeit inconsistently) in present-day quantum mechanics. Thus, whereas most laws describe truly the way things appear to observation, theories approach the goal of giving a true explanation of why things are as they appear to be.

MEANING AS WORD USE

A view of scientific progress closer to the traditional one has been taken by philosophers in the school of Wilfrid Sellars.[18] Gary Gutting, for example, agrees with Hesse that although terms change their meaning from theory to theory, they do not change it entirely, and that consequently theories can be compared. However, he takes the Wittgenstein–Sellars view that the meaning of a term is equivalent to the ways in which the term is used in a language-system. In different theories, he says, the same terms often are used in similar ways and so have similar meanings, even when they refer to different things. Look again at the term "mass" in classical and relativistic mechanics. Both theories define momentum as the product of mass and velocity; both treat mass as a measure of inertia; both relate mass (though in different ways) to kinetic energy; and both measure mass in similar ways (e.g., by scales and mass spectrometers). These similarities in role justify the use of the same term in the two theories.[19]

Gutting then maintains that if terms are used in similar ways in successive theories, the true substance of an earlier theory can be included in a later one. An old truth is transferred to a new theory when the latter contains a true statement which plays a part like that of the corresponding statement in the earlier theory.[20] For example, the ratio of an electron's mass to its charge has the same value in classical physics, special relativity, and quantum mechanics. This is not because J. J. Thomson's classical statement of this truth appears in all three theories (changes in the meaning of "mass" and "charge" make this impossible), but because it has true analogies in the other two theories.

How, then, does science progress through changes in theory? Science progresses, in this view, whenever a new theory explains all that its predecessor explained and more besides. The new theory will reject what has been found false in the old theory, but it will retain, in the form of correspondent statements, what has been found true or approximately true. The new theory also will explain the relative success of the old theory. That is, it will show why phenomena obey the laws of its predecessor to the extent that they have been observed to.

Consider how Newton's mechanics explains the partial success of the laws of Kepler and Galileo. Kepler's laws state that a planet moves round the sun in an ellipse. Newton's theory shows that the path of a planet is not quite an ellipse because a planet is influenced by the perturbational gravitational forces[21] of other objects besides the sun (other planets, satellites, asteroids, interplanetary dust and

35

gas). Galileo's law of falling bodies asserts that a body falling freely near the earth's surface accelerates constantly. Newton's theory shows that the acceleration is not quite constant because the force of gravitation varies with the distance of the falling body from the center of the earth. Nevertheless, Newton's theory shows that Kepler's and Galileo's laws are approximately true, since the perturbational forces acting on a planet, and the relative change in distance between a falling body and the center of the earth, both are near zero.

Another former student of Sellars, Jay Rosenberg, points out that we can sometimes calculate by how much the predictions of an old theory differ from those of its successor. Insofar as this difference decreases through a sequence of theories, we may say that the theories "converge" on one another and in this sense approach the truth. Theoretical progress, he declares, "is marked by the decreasing slack with which successive theories fit increasingly refined data."[22] However, we can claim that successive theories have a direction and make progress only if we can actually measure their "mutual convergence." Since the theories of nonquantitative science yield nonquantitative predictions, they must be regarded only as alternatives to one another and not as successively closer approximations to the truth. However, Rosenberg does not mention any group of theories that actually have converged, nor does he show at all precisely how the calculation is to be done.

METAPHYSICAL BLUEPRINTS

But suppose it can be done. Must we conclude that only quantitative theories can approach the truth? Not at all. The overall aim of science is to discover the order we *assume* is present in nature. If we can state (a) in very general terms what this order must be like (e.g., that it must be simple), (b) what aspects of this order are sought by the several sciences, quantitative and nonquantitative alike, and (c) how our statements of (a) and (b) can be improved over time, we shall have articulated the implicit goals of science in whole or in part. If we then can show that theories in the various sciences have successively become more simple and coherent (and thus more adequate representations of the order they are intended to explain) and at the same time more empirically successful, we shall have shown that science as a whole and also its component sciences have indeed made progress toward their respective goals. This thesis has been proposed by Nicholas Maxwell.[23]

Science, says Maxwell, is guided by the metaphysical presupposition that the universe is intelligible. The fundamental aim of science is to develop increasingly simple and coherent theories that progressively make this conjecture more detailed and more true and at the same time meet with increasing empirical success. Ugly, inconsistent theories, even well-confirmed ones, are in the long run rejected because they postulate only disorder and so do not help us realize this aim. Such is the permanent "metaphysical blueprint" of science, the blueprint for all other, more specific, blueprints.

A mature science, says Maxwell, usually is guided by a metaphysical blueprint or set of assumptions, often implicit, which seeks to specify the invariant features of a domain in very general terms, so that testable theories may be developed which represent them more precisely. Each blueprint, then, specifies certain sorts of fundamental entities and their properties. The blueprint behind classical physics postulates three-dimensional Euclidean space, one-dimensional continuous time, and mass-points interacting through centrally directed forces varying continuously with distance. The Einsteinian blueprint specifies a four-dimensional space-time continuum, with mass-energy giving rise to gravitation as a geometrical property of this continuum.[24]

According to Maxwell, the blueprints of science tend to improve, becoming more intelligible and more empirically successful. Physics, for instance, has been guided by five different blueprints: the Aristotelian blueprint, which states that phenomena occur because things strive to realize their potentials; the blueprint of the corpusculareans and Cartesians, which asserts that all phenomena are to be explained in terms of the motions of corpuscles interacting by contact; the blueprint of classical mechanistic physics (Newton, Boscovich, Helmholtz), which maintains that all phenomena result from motions of mass-points interacting through central forces of attraction and repulsion varying with distance; the blueprint of Einsteinian field theory, which attempts to explain all phenomena as effects of a unified field; and the blueprint of quantum mechanics, still under debate, which proposes that nature is ultimately nondeterminist. This sequence strikingly exemplifies the advance of science.

Scientific progress, then, is not a matter of empirical growth alone. It does not consist solely in inventing theories that predict more phenomena more accurately. It also involves proposing theories of increasing intelligibility, including greater beauty and simplicity. The more intelligible our theories become, the more we understand the intelligibility of the world.

WHAT FUTURE FOR SCIENCE?

Science has been practiced for at least 2,500 years. How long will it continue? For as long, it may be replied, as civilization lasts. Granted, governments can put an end to organized scientific activity whenever they choose. Conceivably they might decide that further scientific inquiry would produce knowledge of grave danger to mankind; or they might conclude that the free spirit of science threatened their security. But that is unlikely. If men are to grow in power to control the conditions of their lives, they must increase their knowledge of nature; and it is hard to imagine governments as a whole declaring that, since men now were absolute masters of their destiny, the quest for scientific knowledge might as well cease. To be sure, a given basic science might appear to be spent. Its theories might have survived all tests, and further experiments might not alter current knowledge. Physicists, for example, might find themselves able to predict the behavior of matter under all conditions they had encountered in the universe and under any they cared to create. But other sciences still would thrive and our interests would continue to expand. Moreover, the growth of such sciences, not to mention that of technology (e.g., space travel), surely would generate new problems for physics to solve.

Nevertheless, is it not possible that one day we shall have learned all there is to be known about man and the world? After all, there is only one universe to investigate; and science, unlike mathe-

38

matics or philosophy, cannot perpetuate itself by inventing purely abstract problems. Biologist Bentley Glass, for one, maintains that we already know the laws of matter and life. Throughout the universe, he says, matter consists of the same particles and elements. The genetic code is universal, and in all species energy is metabolized in similar ways. No more fundamental breakthroughs remain to be made. "We are," he says, "like the explorers of a great continent who have penetrated to its margins in most points of the compass and have mapped the major mountain chains and rivers. There are still innumerable details to fill in, but the endless horizons no longer exist."[25]

I disagree. Take the state of contemporary physics. The two guiding theories in this science, general relativity and quantum mechanics, are mutually inconsistent. According to general relativity, matter is an aspect of space-time, which is continuous and determinist. According to quantum mechanics, matter is discontinuous or particulate (but with wave characteristics) and ultimately nondeterminist. General relativity, moreover, has been under fire from a succession of critics for over half a century. Quantum mechanics is in even worse shape, since it employs the contradictory concepts of wave and particle. Furthermore, it yields predictions about the results of measuring microphenomena only when it is coupled with part of classical physics to describe the macroscopic measuring instrument. Thus the theory in experimental use actually is a conjunction of two partly inconsistent theories, one of which is supposed to have refuted the other! Until a deeper theory has been proposed which unifies quantum mechanics and reconciles it with general relativity, we cannot claim that scientists have discovered the ultimate laws of matter. As for the laws of life, biology is less unified than physics. It is unable to show how the replication of DNA (explained by molecular biology) guides the actual growth and functioning of organisms (explained by biology's other branches).

If and when the problems facing current theories are solved, others undoubtedly will remain. Take the fundamental constants of physics. These are quantities whose numerical values are not explained by the theories in which they appear but are added to the theories after being ascertained empirically. Quantum mechanics cannot explain why the electron has the mass and charge that it does, nor can special relativity account for the speed of light. Or take certain ratios. The number 137.036 relates the size of the electron to the size of the atom, and the size of the atom to the wavelength of the light emitted. Astronomical observation has shown that this ratio holds exactly for atoms even a billion years hence in space and time. Why this number and not another? Physicists do not doubt that other fundamental ratios, such as the ratio of the mass of the proton to that of the electron, are uniform throughout the universe. Are these ratios aspects of an underlying mathematical structure? Paul Dirac, for example, has suggested in his Large Numbers Hypothesis that the really big dimensionless numbers in nature are related to one another and to the age of the universe.[26] Physicists, however, have barely begun to tackle these questions.

But will nature itself exceed our comprehension? The universe may well consist of an infinite variety of things arranged in an endless succession of levels or relatively independent domains. These levels include the human level, the level of animals and plants, the level of inanimate objects (rivers, stones, stars), the level of molecules and atoms, and the level of elementary particles. Above and below these levels we must expect to discover others, such as a level of beings more evolved than ourselves, occupying other galaxies perhaps or other dimensions of the universe, and a level of matter more primitive than the subatomic. Will there be too much for us to discover? The question seems unanswerable. For how shall we ever know whether we have fully explained the natural order? All we shall be able to say is that our most persistent efforts have failed to take us further.

Are there, then, any *a priori* limits to what we can know? Science certainly is limited by the scope of the human mind and senses. There are notions we cannot meaningfully pose for investigation because, as Kant said, our mental categories do not permit them. We cannot, for instance, propose a scientific theory that denies the existence of space and time, for we cannot think of the physical world without these concepts. Again, if the universe is infinite, there are parts we shall never observe. If it is finite, on the other hand, we shall observe most parts only as they used to be. With the aid of the most powerful telescopes we can now see about 8 billion light-years, that is, the distance that light traveling at 186,000 miles per second would travel in 8 billion years, or about 53,000 million million million miles. Yet this distance is the merest fraction of the span attributed by most scientists to a finite universe. However, the mind and the senses evolve, and we cannot assume that our present conceptual and perceptual limitations are permanent. Moreover, we probably have faculties that science has yet to employ, such as the capacity for time-travel with the "astral body" (a body of more subtle matter said to accompany the physical body and to manifest itself as an aura or colored mist). If there are more advanced civilizations elsewhere in the universe, we may be able to communicate with them telepathically or otherwise and eventually create a truly cosmic science.[27]

In principle, then, there seem to be no limits to the advance of science. But are there likely to be limits in practice? Will science come to an end, not because it has attained its goal, but because it has lost the will to do so? Eugene Wigner suggests that formulating ever deeper theories to explain successive layers of nature may in the end be too much for man's intellect and motivation.[28] (Similarly, the English biologist J. B. S. Haldane once said that the universe is not only a good deal "queerer" than we know, it is a good deal queerer than we *can* know.) As thinking becomes more specialized and more remote from experience, says Wigner, it becomes more difficult and appeals to fewer people. The time may come when new students in some branch of science, such as physics, may lack the interest or even the ability to think beyond the level of abstraction of the current theory.

To this I reply that in culturally advanced areas abstractions soon are assimilated, and that so long as physics interacts with technology, it is unlikely to become either uninteresting or altogether remote from experience. Both Newtonian mechanics and Einsteinian relativity were enormously difficult when they first appeared. Yet within a century Newton's theory was common knowledge to scientists with a strong mathematical background, and in less than half that time the principles of relativity were being taught to undergraduates. It seems highly unlikely that nature, which has raised questions and stimulated answers in some of the most powerful minds from the time of the Pre-Socratics, will ever cease to do so. On the other hand, the questions may become more difficult, and, for a period at least, fewer such minds may respond to them than during the past 100 years. If so, science may grow more slowly.

Has science in fact already begun to do so? Some people think so. Derek de Solla Price has studied the rate of scientific growth since 1750, using as indicators numbers of scientists, scientific journals, and scientific abstracts.[29] He concludes that until 1950 both science as a whole and its individual branches grew exponentially, that is, in proportion to their current size. The rate of exponential growth can usefully be described by reference to the time needed for the quantity to double. According to Price, science has doubled every 10 to 15 years. Important findings, however, as opposed to information in the gross, have doubled every 30 years. This is because scientific knowledge, being cumulative, grows like a pyramid, and must double three times in volume for every doubling of height.

But, as Price points out, science cannot expand at this rate indefinitely, for any growth in scientific information depends on a proportional increase in the number of scientists and the amount of money spent on science, both of which are limited. If science continued on an exponential curve until the middle of the next century, "we should have two scientists for every man, woman, child, and dog in the population, and we should spend on them twice as much money as we had."[30] So science must grow more slowly. Instead of climbing ever more steeply, its curve will flatten, perhaps with marked fluctuations, eventually becoming S-shaped (or what is called a "logistic" curve). According to Price, the curve began to ease around 1950.

Price's argument can be faulted in detail. He derives the doubling period of science from the number of abstracts, ignoring work published before the start of abstracting services. If this work is taken into account, the doubling period is more like 28 years.[31] Moreover, science may well renew itself and surge forward again, vaulting the saturation point of the logistic curve.[32] As Price himself has shown, the number of universities in Europe grew exponentially from 950 to 1460, then logistically, and then, with the adoption of a new form of university around 1610, exponentially again.[33] Why should science not do likewise? Again, contrary to Price, science seems to have grown faster since 1950. If we measure growth by the number of science doctorates granted per year in the United States, we find that the growth rate rose from about 3 percent in 1950 to nearly 11 percent in 1969.[34] Finally, *basic* knowledge may well grow much more slowly than we had believed. Between 1855 and 1955, for example, the rate of discovery of laws thought important enough to be named after their discoverers (the practice of eponymy) remained constant at an average of two per year.[35]

Although Price's timing may be wrong, his general thesis—that the rate of scientific growth has been exponential and may soon decrease—surely is sound. The older sciences especially are likely to grow more slowly, and their practitioners may become less convinced that they are advancing steadily toward an explanation of all phenomena in their domains. Most science has been done in the past 400 years; to this extent science is much younger than art and philosophy. Modern artists and philosophers inherit the achievements of many centuries and cultures; they have a multiplicity of styles and precedents by which to assess their attempts; they have many goals rather than a single goal toward which they are inevitably bound. It can be argued that modern physics, with its range of specialties, its group of largely worked-out branches (classical mechanics, optics, thermodynamics, electromagnetism), and its increasing theoretical disagreements (over the foundations of quantum mechanics or alternatives to general relativity), may be approaching this state. The pluralism of physics, which the other natural sciences seem likely to imitate, is more probably the condition of a mature intellectual discipline than that of a spent science.[36]

SUMMARY

Although science replaces many of its conclusions, in the long run it advances closer to the truth of nature. This relatively traditional view of the progressiveness of science has been upheld by Popper. It has been criticized, however, by Kuhn and Feyerabend, who maintain

43

that since scientific terms change meaning from one theory to the next, theories cannot be compared, so that there is no way of telling whether one theory is closer to the truth than another. This criticism can be met by adopting the view that the meaning of a term is its intensional reference (Hesse) or a function of the term's use in a context (Sellars). If theories can be compared in certain respects, then the true substance of an old theory can be expressed in analogous statements in the theory replacing it. However, as Maxwell points out, science does not advance solely by increasing the empirical content of its theories. In order to better explain the order of nature, which is, he says, both simple and beautiful, science must propose theories that are themselves more simple and beautiful.

How long will science continue to grow? Since the levels of nature appear endless, there may be no limits to what science can discover. True, by the time science has explained the nth level, it may have become so abstract that no one will have the desire or even the capacity to pursue it further. Yet, as history shows, what appears impossibly abstract to one generation often is much less so to the next. Still, it seems clear that as science probes deeper into the universe, the task of explanation will become more difficult. Thus science may well grow more slowly, though without coming to a stop.

Science, then, grows toward the truth but may do so less rapidly. Does it also grow rationally? Let us see.

NOTES

1. Geometrical optics was founded by Greek scientists, notably Euclid and Ptolemy, who discovered, among other things, that the angle of reflection of rays from a surface is equal to the angle of incidence.

2. According to this theory, popular in the eighteenth century, particles of the same element are held together by forces of mutual affinity.

3. Thomas S. Kuhn, *The Structure of Scientific Revolutions*, pp. 130–35.

4. Larry Laudan, "Two Dogmas of Methodology," *Philosophy of Science* 43 (December 1976): 589–90.

5. The caloric theory of heat, widely held in the eighteenth century, asserted that heat was a weightless fluid (called "caloric"). In the latter half of the nineteenth century this theory was replaced by the theory of statistical mechanics, which states that heat is caused by the motions of the particles of which bodies are composed. During the eighteenth century, but in the phlogiston, believed that material bodies contained a weightless vapor, termed "phlogiston," which the bodies gave off when heated. At the end of the century, Antoine Lavoisier proved experimentally that no such vapor exists.

6. Popper proposes the notion of verisimilitude in *Conjectures and Refutations*, ch. 10 and addenda; and in *Objective Knowledge: An Evolutionary Approach* (Oxford: Clarendon, 1972), chs. 2, 9.

7. We shall consider Popper on theory-testing in the next chapter.

8. See the essays by Pavel Tichý, John H. Harris, and David Miller in *British Journal for the Philosophy of Science* 25 (June 1974) (hereafter cited as *BJPS*); by Tichý in *BJPS* 27 (June 1976): 25–42; by Miller in *BJPS* 27 (December 1976): 363–81; and by Adolf Grünbaum, "Is the Method of Bold Conjectures and Attempted Refutations *Justifiably* the Method of Science?" *BJPS* 27 (June 1976): 105–36. See also Popper's reply, "A Note on Verisimilitude," *BJPS* 27 (June 1976): 147–59.

9. Thomas S. Kuhn, "Reflections on My Critics," in *Criticism and the Growth of Knowledge*, ed. Imre Lakatos and Alan Musgrave, pp. 266–67.

10. Feyerabend, "How to Be a Good Empiricist: A Plea for Tolerance in Matters Epistemological," in *Philosophy of Science, The Delaware Seminar*, ed. Bernard H. Baumrin (New York: Wiley, 1963), 2:14: "That the relativistic concept and the classical concept of mass are very different becomes clear if we . . . consider that the former is a *relation*, involving relative velocities between an object and a coordinate system, whereas the latter is a *property* of the object itself and independent of its behavior in coordinate systems."

11. Kuhn, *Structure of Scientific Revolutions*, p. 101: "Though the formulas of Newton's Laws are a special case of the laws of relativistic mechanics, they are not Newton's Laws. Or at least they are not unless those laws are reinterpreted in a way that would have been impossible until after Einstein's work."

12. *Ibid.*, ch. 13 and pp. 205–7.

13. Mary Hesse, *The Structure of Scientific Inference* (Berkeley and Los Angeles: University of California Press, 1974), pp. 61–66. The meaning-variance thesis is dis-

cussed by Dudley S. Shapere, "Meaning and Scientific Change," in *Mind and Cosmos*, ed. Robert G. Colodny (Pittsburgh: University of Pittsburgh Press, 1966), pp. 41–85; and by Frederick Suppe in *The Structure of Scientific Theories*, ed. Frederick Suppe, pp. 199 ff. The debate is summarized in Arthur Fine, "How to Compare Theories: Reference and Change," *Noûs* 9 (March 1975): 17–32.

14. Hesse, *Structure of Scientific Inference*, pp. 299–300.

15. Mary Hesse, "Reasons and Evaluations in the History of Science," in *Changing Perspectives in the History of Science*, ed. Mikuláš Teich and Robert Young, p. 147.

16. Hesse, *Structure of Scientific Inference*, pp. 299–301, and "In Defense of Objectivity," *Proceedings of the British Academy* 58 (1972) (London: Oxford University Press [for the British Academy], 1974), pp. 281–84.

17. A. S. Eve, *Rutherford* (New York: Macmillan, 1939), p. 384.

18. Sellars's main works are *Science, Perception, and Reality* (New York: Humanities, 1963), and *Science and Metaphysics* (New York: Humanities, 1968).

19. Gary Gutting, "Conceptual Structures and Scientific Change," *Studies in History and Philosophy of Science* 4 (November 1973): 223–26.

20. This point is treated more fully in Sellars, *Science and Metaphysics*, ch. 5.

21. A perturbation is an influence so small that the problem into which it enters can be solved much more simply if the perturbation is neglected.

22. Jay F. Rosenberg, *Linguistic Representation* (Dordrecht, Holland: Reidel, 1974), p. 94. See also his "The Elusiveness of Categories, the Archimedean Dilemma, and the Nature of Man: A Study in Sellarsian Metaphysics," in *Action, Knowledge, and Reality: Critical Studies in Honor of Wilfrid Sellars*, ed. Hector-Neri Castañeda (Indianapolis: Bobbs-Merrill, 1975), pp. 171–76.

23. Nicholas Maxwell, "The Rationality of Scientific Discovery, Part I: The Traditional Rationality Problem"; and "The Rationality of Scientific Discovery, Part II: An Aim-Oriented Theory of Scientific Discovery."

24. Maxwell, "The Rationality of Scientific Discovery, Part II," p. 275.

25. Bentley Glass, "Science: Endless Horizons or Golden Age?" *Science* 171 (January 8, 1971): 24.

26. Paul Dirac, "The Development of the Physicist's Conception of Nature," in *The Physicist's Conception of Nature*, ed. Jagdish Mehra (Dordrecht, Holland: Reidel, 1974).

27. On this topic see *Communication with Extraterrestrial Intelligence*, ed. Carl Sagan (Cambridge: MIT Press, 1973). See also Sagan's latest work, *The Dragons of Eden: Speculations on the Evolution of Human Intelligence* (New York: Random House, 1977), in which he declares that men must reach out to other intelligent beings in the universe and unite human history with that of the cosmos itself.

28. Eugene P. Wigner, *Symmetries and Reflections: Scientific Essays* (Bloomington: Indiana University Press, 1967), pp. 215–16. Wigner suggests, however, that, by pooling intellects, team science may possibly accomplish what the individual scientist cannot. See also Alvin M. Weinberg, *Reflections on Big Science* (Cambridge: MIT Press, 1967), pp. 44–45.

29. Derek J. de Solla Price, *Little Science, Big Science* (New York: Columbia University Press, 1963), and *Science Since Babylon* (New Haven: Yale University Press, 1961), ch. 5.

30. Price, *Little Science, Big Science*, p. 19.

31. K. O. May, "Quantitative Growth of the Mathematical Literature," *Science* 154 (December 1966): 1672.

32. Steven Rose, "The S Curve Considered," *Technology and Society* 4 (1967): 33–39.

33. Price, *Science Since Babylon*, pp. 114–15.

34. U.S. Bureau of the Census, *Statistical Abstracts of the United States, 1971* (Washington, D.C.: Government Printing Office, 1971). Cited by G. Nigel Gilbert and Steve Woolgar, "The Quantitative Study of Science: An Examination of the Literature" (essay review), *Science Studies* 4 (July 1974): 282.

35. Donald de B. Beaver, "Reflections on the Natural History of Eponymy and Scientific Law," *Social Studies of Science* 6 (February 1976): 94.

36. See Paul M. Quay, "Progress as a Demarcation Criterion for the Sciences," *Philosophy of Science* 41 (June 1974): 158–69.

Chapter 3

From Conjectures to Paradigms

SCIENTIFIC RATIONALITY

Science evolves through the acts of men and women—acts such as inventing hypotheses, conducting experiments, weighing evidence, and publishing results. The purpose of these acts is to produce verified knowledge—knowledge that is worthy of acceptance by the scientific community. In order to produce such knowledge, science must be rational. For unless claims to knowledge are rationally based, there are no grounds for preferring them to the claims of gurus and fortune tellers, and the scientific quest is pointless. If we are to understand the scientific enterprise, then, we must find out not only how science evolves but also how rationally it does so.

What is it to be rational? Philosophers have answered this question in many ways. But the core of rationality seems to consist in believing or doing things for good reasons. In science an act is rational in part if the scientist has good reason to believe that it is the one most likely to achieve a goal of science. Such an act is the best means to an existing goal. But rationality is not limited to means. The act is fully rational only if the scientist has good reason to believe that the goal is the best one for science under the circumstances. Nevertheless, he may be mistaken or self-interested. Therefore, science as a whole cannot be rational unless its goals, and also its means, are continuously evaluated by all scientists. For science is a collective enterprise, and without collective debate there is no way of judging whether current goals are the best for science as a whole or only for certain scientists.

That scientific rationality requires collective debate aimed at collective agreement also can be seen in the problem-solving activity of science. The solution to a scientific problem is a logical argument in which conclusions are drawn from premises, and evidence is supplied to show that the conclusions are in fact true. The solution is appraised by examining whether the conclusion is implied by the premises and supported by the evidence. Clearly, if a solution is proposed as both logically correct and factually true, it must be verifiable by all qualified scientists, which means, at least, by all scientists working in the field. This is not to say that they will reach agreement straightaway (it may take them a decade, a generation, or longer) but rather that science cannot be rational and yet not seek their agreement.

To be rational, then, science must seek universal agreement, at least within the discipline or specialty. Nevertheless, universality in this sense is only the *form* of scientific rationality, not the substance, for people can be collectively irrational, and scientists may think they act rationally when they do not. A unanimous decision by astronomers to return to astrology would not necessarily be rational. Collective agreement is a necessary but not a sufficient condition for rationality in science.

What, then, is the *substance* of scientific rationality? Most philosophers of science agree that the general aim of science is to increase our stock of empirical knowledge, that is, to propose theories that successfully predict more facts than their predecessors. But they disagree over how this aim should be achieved. Some favor theories that logically entail prior theories. Some contend that scientists should invent bold, speculative theories and test them ruthlessly. Some argue that scientists should work out the implications of a single theory and uncover all the facts it can predict. Some, again, call for competition between research programs to see which program can successfully predict most facts. Yet again, some urge scientists to propose and test as many theories as possible, especially those that contradict established theories. Other philosophers still, with whom I agree, deny that empirical growth is a sufficient end for science. They contend that science should seek theories that not only are more empirically successful than their predecessors, but also represent the order of nature more simply, coherently, and aesthetically.

Before discussing these views, however, I must emphasize that scientists act in response not only to scientific findings but also to fac-

49

tors outside science altogether, such as philosophic theories, religious doctrines, public opinion, government decisions, and the availability of research funds. Nor do these factors necessarily promote irrational behavior. Newton, for instance, was not irrational when he argued that space is empty so that God can fill it with His presence, for in a religious society the aim of science was to prove the existence of God from the evidence of nature. On the other hand, scientists can be irrational even when acting from solely scientific considerations. The British physicists who dismissed Faraday's ideas of the electromagnetic field had minds closed by their Newtonian training. As a result of their indifference Faraday broke down and died early, and the advance of physics was delayed for a generation. I shall examine the influence of social and cultural factors later in the book. For the moment I will consider the nature of scientific change from a perspective within the scientific movement itself.

HUME ON INDUCTION

The current debate on scientific change began as a reaction to the philosophy of logical empiricism. Let us therefore look at the problem of induction which the logical empiricists sought to solve. This is the problem of what grounds we have for reasoning from particular instances to general statements. The laws of science have the form "All A's are B's," e.g., "All pieces of copper expand on heating." The laws are accepted either because a great many examined A's have been B's and no examined A's have failed to be B's, or because the laws have been deduced from other laws that have been established in just this way. Thus the law cited was originally proposed, and has been justified since, on the grounds that all pieces of copper that have been heated have been found to expand. The logical operation of making or justifying an unrestrictedly general statement on the strength of observing a number of particular intances is called "induction."

But as David Hume convincingly showed, no number of observation statements ever logically entail a general statement. His reason was simple but devastating: It is never contradictory to assert, "All observed A's are B's, but some unobserved A's are not." Indeed, we cannot even deduce that the next observed A will be a B, for it always is logically possible that it will not be. Although the sun may have risen every day since the world began, it does not have to rise tomorrow. Nor can one prove that induction is reliable by arguing that most past inductions have been so, for it can be replied without contradiction that the next induction may not be. Yet all scientific laws and theories take for granted that the future will resemble the past, that what has been observed hitherto will be observed hereafter. Thus the assumption that nature is regular presupposes the validity of induction. Therefore, concluded Hume, since induction may not be valid, science may not be valid. Its laws and theories have no rational guarantee, since they can always be refuted by the next instance.

LOGICAL EMPIRICISM

The logical empiricists, successors to the logical positivists of the 1920s, answered Hume as follows: We may not be able to prove for certain that a scientific law or theory is true, but we can calculate the *probability* that it is. We do so by working out the ratio between (a) the number of predictions that have been derived from the theory and have been confirmed, and (b) the total number of predictions that can be derived from the theory. If this ratio is high enough, the theory can be considered well confirmed. Some logical empiricists, such as Rudolf Carnap, have proposed logical principles for calculating such ratios and the degree of probability they give to a theory.[1] For logical empiricism, then, inductive science *is* rational because, in principle, at least, it can tell the scientist how likely his theories are to be true and hence how far he can trust them.

However, as more data are acquired, a new theory may be proposed that agrees better with the evidence than the established theory. Do we then replace the old theory with the new one, or do we retain them both?

We retain them both, say the logical empiricists. In contrast to most nineteenth-century writers, who had held that science changes mainly by adding new laws and observation reports to old ones, the logical empiricists assumed that science grows, and grows rationally, by turning established laws and theories into logical consequences of new, more comprehensive theories. This process is called "reduction," and the old or "reduced" theory is said to be a "special" or "limiting" case of the new one. In the words of Ernest Nagel,

> The phenomenon of a relatively autonomous theory becoming absorbed by, or reduced to, some other more inclusive theory is an undeniable and recurrent feature of the history of modern science. There is every reason to believe that such reduction will continue to take place in the future.[2]

For instance, Newton's theory of mechanics and gravitation explained various existing laws of motion, such as Galileo's law for bodies falling freely near the earth's surface and Kepler's laws of planetary motion. Whereas these laws described the motions of bodies in limited regions hitherto regarded as distinct (terrestrial and celestial space), Newton's theory described motion throughout space. Later, Newton's laws of motion were reduced to Einstein's special theory of relativity and his law of gravitation to the general theory. Similarly, according to logical empiricism, thermodynamics was reduced to statistical mechanics, and many chemical laws were explained by quantum mechanics. Today attempts are being made to reduce classical genetics to molecular biology.

As the logical empiricists saw it, new scientific theories generally are proposed to explain old ones, that is, to show that the latter hold for a limited range of phenomena and are logically entailed by theories that are more comprehensive. So-called revolutionary theories, like relativity and quantum mechanics, do not refute and replace established theories; they simply set limits to them. As a rule, said the logical empiricists, to be acceptable a new theory must either entail the existing theory or at least be consistent with it. More specifically, if T and T' are two theories—T' the theory to be explained, and T the explaining theory—then T' is explained by being deduced from T together with statements of initial conditions specifying the phenomena for which T' is true.[3]

52

CRITIQUE OF LOGICAL EMPIRICISM

The logical empiricist position is open to serious objections, and as a result it has lost most of its support among philosophers of science. First, the logical empiricists claim that science is rational insofar as it is inductive, yet most of them have had to admit that their theories of probability and confirmation are generally inapplicable. Carnap, for instance, concedes that his inductive logic fails to explain any important episode in the history of science, and notes that on his reckoning not one universal scientific theory is in the least confirmed.[4] But if we cannot say how probable our theories are, we cannot state the odds against their refutation by the next instance. Thus the logical empiricist case against Hume collapses.

Again, the logical empiricists argue for reduction, yet, as we have seen, owing to changes in the meanings of terms one theory rarely can be reduced to another. I do not deny that sentences analogous to some of those in a predecessor theory usually can be deduced from the premises of its successor. But the earlier theory is reduced neither logically nor in its entirety. Moreover, as Feyerabend has pointed out, if we require every subsequent theory to include its predecessor as a special case, we are obliged to dismiss any new theory that is logically inconsistent with the prevailing one. How rational would we be to reject Einsteinian relativity, which successfully predicts the Newtonian facts and more besides, simply because it contradicts the theory that was accepted first?

The logical empiricists were more interested in the structure of scientific knowledge than in the process by which that knowledge is changed. They paid little attention to the ways in which theories are extended and criticized and none at all to the process of invention. Theories, they held, are created by genius, or intuition, or accident, not by rational thinking. They are inspired guesses that cannot be made rationally but can be tested rationally (see chapter 5). But they asserted this as a dogma without bothering to verify it. In sum, logical empiricism was wide open to attack by any philosopher with an interest in scientific change.

53

KARL POPPER

The first attack on logical empiricism came from Karl Popper. Once a member of the logical positivist Vienna Circle, Popper argued that the logical empiricist solution to the problem of induction is no solution at all. However probable a theory may seem, it still can be refuted by the next piece of evidence to turn up. (No matter how many A's are found to be B's, we cannot be sure that some remaining A's are not B's.) But if we cannot verify a theory, perhaps we can falsify it. No number of statements reporting the observation of A's that are B's logically entails the generalization "all A's are B's." But a single statement reporting the observation of an A that is a C does entail the statement "not all A's are B's." It follows that *although a theory can never be conclusively proved, it can be conclusively falsified.* Hence the aim of science as a rational enterprise is to invent theories that are falsifiable and to test them by seeking to refute them.[5]

For Popper, then, science advances by proposing audacious theories ("conjectures"), by making every effort to falsify them ("refutations"), and by accepting provisionally only those theories that survive this process. In his own words, "There is no more rational procedure than the method of trial and error—of conjecture and refutation; of boldly proposing theories; of trying our best to show that these are erroneous; and of accepting them tentatively if our critical efforts are successful."[6] Popper praises the bold, falsifiable theories of Newton, Maxwell, and Einstein, and the "crucial" experiments carried out by Michelson and Morley to test the ether hypothesis, and by Eddington to test the general theory of relativity—crucial because they confirmed one hypothesis and disconfirmed others. He concludes that whenever science protects its theories, it stagnates.

Let us look more closely at Popper's triad of conjectures, refutations, and tentative acceptances. The more a theory purports to explain, says Popper, the more open it is to refutation, since it yields more predictions that can be disconfirmed. In this sense, highly falsifiable theories are "highly improbable" before being tested. When theories compete, the least probable should be tested first, for if it survives, we can have more confidence in it than we could have had in the others. Since the aim of science is refutation, not confirmation, everything is to be gained by exposing, not concealing, the weaknesses of a theory. The scientist should state in advance those tests that he regards as potential refutations of his theory. In particular, he should make "risky predictions" asserting the existence of states of affairs not deducible from previous knowledge and if possible excluded by it, states of affairs which only the theory itself leads us to expect. "Confirmations should count," he says, "only if they are the result of *risky predictions*; that is to say, if, unenlightened by the theory in question, we should have expected an event which was incompatible with the theory—an event which would have refuted the theory."[7]

According to Popper, a theory normally is falsified when a prediction derived from it disagrees with an observed event in what supporters and critics alike agree is a serious test.[8] A theory that cannot be falsified—and for Popper the prime examples are Marxism and psychoanalysis—is not scientific but metaphysical or dogmatic. "Irrefutability is not a virtue of a theory (as people often think) but a vice."[9] A refuted theory is a success, not only because the theory may possess intellectual beauty, but because it leads to the discovery of the fact which refutes it and so stimulates research to explain it.

But it is not enough to falsify theories. At least some theories must resist falsification and in this sense be "corroborated." Only corroboration tells us which of our theories describe the actual world and provide fruitful leads for further research. A run of swiftly refuted theories would bring research to a dead end, because it would leave scientists in the dark about the relation of their theories to the world and about which hypotheses were worth trying.

For Popper, finally, all good science is revolutionary. Science grows through the continual overthrow of theories by refuting facts and through their replacement by theories that explain more facts. In his own words:

The theories of Kepler and Galileo were unified and superseded by Newton's logically stronger and better testable theory, and similarly Fresnel's and Faraday's by Maxwell's. Newton's theory, and Maxwell's in their turn, were unified and superseded by Einstein's. In each such

55

case the progress was towards a more informative and therefore logically less probable theory: towards a theory which was more severely testable because it made predictions which, in a purely logical sense, were more easily refutable.[10]

CRITIQUE OF POPPER

The heart of Popper's philosophy is the liberating idea that science grows through adventurous, falsifiable theories—an idea which biologist Peter Medawar and neurologist David Eccles both declare has inspired their own research. Nevertheless, the idea is only partly true, and in other respects Popper's theory of science is seriously flawed.

Although Popper claims to have rid science of induction, he reintroduces the notion in his account of corroboration and test statements. To say that we can have more confidence in a theory that has withstood our determined efforts to refute it is in effect to claim that we can be more confident now because our previous confidence was justified. But this is inductive reasoning. Again, we verify a test statement by repeating the observation it records. But to accept the first test on the strength of the second is to accept it inductively.[11] Thus Popper relies on induction after all.

Popper maintains that all good science is revolutionary. But as Kuhn and others have pointed out, science also grows by systematically extending a major theory through the solution of the problems it raises. Instead of seeking refutations, scientists uncover more facts by applying the theory to the many different kinds of situations it promises to explain. Popper ignores this part of science.

Again, if scientists had followed Popper to the letter, science would have lost some of its most successful theories. In fact, apparent refutations often are and have been ignored in the hope that they will prove inaccurate.[12] Galileo, for instance, promoted the Copernican theory in the face of what at the time seemed overwhelming contrary evidence. Newton's theory was retained in the face of such anomalous facts as the precession of Mercury (footnote 15). The special theory of relativity was upheld over the contrary evidence of D. C. Miller, who repeated the Michelson-Morley experiment.

56

In my view, falsificationism is not only historically false, but nonrational. Counterinstances often challenge scientists to develop a theory to its full capacity. A promising new theory should not be rejected at the first serious refutation. The theory may have been refuted simply because it was applied incorrectly. In any case, as Lakatos points out, all new theories are "born refuted." One reason for this, as Feyerabend indicates, is that the available evidence has been gathered by scientists guided by theories that support the theory under challenge and hence rest on assumptions antagonistic to those of the new theory. If the new theory is to have a fair hearing, it must be allowed to show how many *new* facts it can predict correctly.

In sum, Popper's theory is a brilliant one-sided conjecture that has been partly refuted but has, in the fashion he himself prescribes, stimulated the proposal of other theories with more empirical content. The first of these theories was advanced by Thomas S. Kuhn.

THOMAS S. KUHN

Where Popper attributes the growth of science to the imaginative and critical activity of the individual scientist guided by the falsificationist ideal, Kuhn argues that science advances when scientists are trained in a common intellectual tradition and use that tradition to solve the problems it raises. Kuhn sees the history of a "mature" science as essentially one of a succession of traditions, each with its own theory and methods of research, each guiding a community of scientists for a period of time, and each in the end abandoned.[13]

Kuhn began by calling the ideas of a scientific tradition a "paradigm" but he now calls them a "disciplinary matrix." He never clearly defines a paradigm, but we may regard it as a world view expressed in a theory.[14] The theory purports to explain the behavior of the basic entities in some sector of the world. The paradigm as a whole determines what problems are investigated, what data are considered relevant, what techniques of investigation are used, and what types of solution are admitted. For example, under the Newtonian paradigm, solutions are achieved in terms of forces and the motions of particles.

Kuhn's disciplinary matrix is conceived more precisely. It consists of four elements: symbolic generalizations, metaphysical assumptions, values, and concrete problem-solutions ("exemplars"). Symbolic generalizations are less laws than "law-sketches"; they yield different, specific laws when applied to different situations. The law $F = ma$, for instance, applies to many situations and is found in many types of equations. Physicists study this formula and learn to look at mechanical problems in terms of forces, masses, and accelerations. Symbolic generalizations are interpreted with the aid of metaphysical assumptions—assumptions that cannot be tested empirically at the time—such as (in the nineteenth century) the assumption that atoms and fields of force exist. Values are the qualities prized in a theory, such as internal consistency, predictive power, and fertility in suggesting problems. Exemplars are model problem-solutions that serve as guides for solving real problems. For instance, Galileo calculated the motion of a ball down an inclined plane by regarding it as akin to the motion of a pendulum, in that (in the absence of friction) the ball gathers sufficient velocity to return to the same height on any second incline. This solution was used as a model by later scientists working in mechanics. With the aid of exemplars, symbolic generalizations such as $F = ma$ are applied to particular types of situations such as freely falling bodies, pendulums, and springs.

Kuhn divides the history of mature, or paradigm-guided, science into alternating "normal" and "revolutionary" phases. During normal science, researchers develop the implications of a paradigm or disciplinary matrix as fully as possible. They neither criticize the paradigm nor seek alternatives to it. The paradigm supplies problems and assures scientists that each problem is soluble. An unsolved problem reflects on the scientist rather than on the paradigm. Throughout the nineteenth century, for example, Mercury's precession was regarded as a challenge to scientists rather than as a falsifier of the Newtonian paradigm.[15]

Revolutions such as those of Copernicus, Newton, Darwin, and Einstein are infrequent, says Kuhn, and they are sparked by crises. A crisis occurs when scientists are unable to solve many long-standing problems confronting a paradigm. The backlog of anomalies then is considered a "scandal," and scientists begin both to test the paradigm and to seek alternatives based on different metaphysical assumptions. Eventually one alternative wins the support of most scientists in the field and is accepted as the new paradigm. Previous knowledge is reconceived or discarded; textbooks are rewritten; courses are altered; and scientists look at the world differently.[16] Kuhn cites the crisis in quantum physics after the breakdown of Bohr's "old quantum theory" in 1922, when rival theories proliferated, notably those of de Broglie, Bohr-Kramers-Slater, Schrödinger, and Heisenberg. The crisis was resolved in 1926 when Max Born showed that the Schrödinger and Heisenberg theories were mathematically equivalent.[17]

Kuhn defends the rationality of normal science on two grounds. First, it is a highly efficient way of solving problems and extending a major theory.[18] The paradigm keeps scientists from arguing endlessly over fundamental assumptions, taking on unproductive or insoluble problems, and disputing with mavericks and cranks. "Immature" sciences, as Kuhn calls them, such as psychology and sociology, make little progress because they lack paradigms; they are divided by warring schools, none of whose members will accept the work of other schools as a basis on which the science as a whole can proceed.

Second, every paradigm paves the way for its successor. Since any theory is an abstraction from reality, no theory can hope to account for all the phenomena in its domain. Sooner or later every theory encounters anomalous facts. Research under a paradigm guarantees that as many of these facts as possible are encountered as soon as possible. When the paradigm is largely worked out and is confronted by the anomalies that mark its limits, scientists look for a successor.

Is revolutionary science rational? Only up to a point, says Kuhn, for the new theory and the old are incommensurable in two respects. First, as I have mentioned, they use some of the same terms in different senses. Second, supporters of the two theories, looking at the world through different exemplars, will observe different facts. Where Aristotelians saw a heavy body constrained from falling, Galileo saw a pendulum. Where Priestley had seen dephlogisticated air and others had seen nothing, Lavoisier saw oxygen.[19]

Nevertheless, between the two groups there is always "partial communication." Each group can translate the problematic terms of the other into the everyday vocabulary of both groups.[20] Moreover, there are standards of theory-comparison that transcend paradigms. They include problem-solving capacity, explanatory power, predictive power, simplicity, internal consistency, and consistency with accepted theories.[21] However, these standards operate as values, being differently prized by different scientists. Thus, while one scientist may admire the prevailing theory for its simplicity, another may support a rival theory because of its predictive power. Scientists, then, may agree that certain qualities are valuable in theories yet disagree on the value of particular theories.

Because the standards of theory-appraisal are values that can be variously applied, scientists cannot prove logically that one theory is better than another. Instead, they must "persuade" one another rationally. Kuhn distinguishes between persuasion and conversion. A scientist is persuaded by a new theory when he judges intellectually that it is superior to the established theory, even though he may not be emotionally sympathetic to it; he is converted to the theory when he finds himself at home in it and sees the world in its terms. A scientist may be persuaded by a theory yet not converted. This was the experience of many scientists who first met with relativity theory and quantum mechanics in middle age.[22] Nevertheless, the first supporters of a new theory commit themselves, as a rule, out of a feeling for the promise of the theory that they cannot fully articulate. In Kuhn's words:

> Something must make at least a few scientists feel that the new proposal is on the right track, and sometimes it is only personal and inarticulate aesthetic considerations that can do that. Men have been converted by them at times when most of the articulate technical arguments pointed the other way. . . . If a paradigm is ever to triumph it must gain some first supporters, men who will develop it to the point where hardheaded arguments can be produced and multiplied.[23]

60

CRITIQUE OF KUHN

Kuhn's account of scientific change is comprehensive and illustrated with a wealth of historical examples. He is the only philosopher so far to have related the intellectual drive of science to the scientific community, and he was the first to stress the persistence of research traditions even in the face of serious anomalies. Nevertheless, his work has certain shortcomings which we must now consider.[24]

Because Kuhn fails to specify its constituents (theory, techniques, standards, etc.), the idea of a paradigm is of little use as a research tool. The concept of a disciplinary matrix is more precise but lacks the fertile, unifying component of a world view. The disciplinary matrix is an assembly of separate parts that seem insufficient to inspire a research tradition over a substantial period of time.

Kuhn proposes normal science, generating intermittent revolutions, as the sole mode of growth for a mature science. Yet science also grows in other ways. Sometimes a major new theory is proposed, not in response to a build-up of anomalies, but to resolve a conflict that has emerged between two existing theories. Einstein introduced his special theory of relativity to reconcile Newtonian mechanics with Maxwell's electrodynamics, and his general theory of relativity to unite the special theory with Newton's theory of gravitation. Not infrequently a fresh theory is proposed to explain a new area of ignorance, as after the discovery of pulsars in 1967.[25] Sometimes a theory may be invented by a group of scientists entering an established discipline from outside, like the phage group, founders of molecular biology, who switched from physics to biology in the late 1930s and early 1940s. A research tradition also may alter its own assumptions, as Mendelian genetics has done in this century.[26] Again, a tradition may be thrown into a state of near crisis by purely theoretical controversy. Both general relativity and quantum mechanics currently are under attack for theoretical reasons, not because they have accumulated anomalies. Moreover, Kuhn does not explain why anomalies, which are always there, will sometimes precipitate the search for a new theory and sometimes not.

Then, again, criticism of a prevailing theory, work on alternative theories, and debate over fundamentals seem to occur in all periods, only intensifying in those that Kuhn calls revolutionary. During the nineteenth century, for example, the Newtonian tradition influenced thinking in most branches of physics but did not dominate it. In the study of electromagnetism two traditions competed—the Continental action-at-a-distance tradition, springing from Ampère and Coulomb, and the British field tradition started by Faraday. Different scientists working in these traditions drew on different Newtonian assumptions. Both traditions were variously compatible with the Newtonian one, yet neither was directed by it. Look at science today. Particle physics is divided between S-matrix theory and quantum field theory. General relativity has been assailed for half a century by such critics as E. A. Milne, Henri Poincaré, Alfred North Whitehead, Fred Hoyle, and Robert Dicke. Today, it is in competition with a range of theories.[27] In quantum mechanics, the Bohr-Heisenberg-Dirac interpretation of quantum theory has been criticized by a succession of scientists such as Einstein, Schrödinger, de Broglie, David Bohm, and Alfred Landé. In biology, the synthetic theory of evolution faces competition from the "neutralist" theory.[28] In sum, in most branches of science there often is a prevailing, or particularly authoritative, theory, but it is seldom without alternatives. Granted, something corresponds to Kuhn's normal science, but it is a far more fluid state of affairs than Kuhn would have us believe.

Strict normal science generally is not very rational, for in it a single theory is extended but not criticized. This limits both the range of inquiry, since alternative theories are not considered, and also the rate of growth, since a Kuhnian theory stimulates the search for a successor only after it has accumulated anomalies. I do not deny that there may be a case for concentrating resources on a single theory on certain occasions. If the dominant theory of a science or specialty is outpacing all competitors, and steadily and successfully predicting new facts, it makes sense to back the theory with men and money until it begins to show diminishing returns, at which point some resources might be diverted to promising alternatives.

Turn now to theory-choice. Kuhn has been accused of claiming that a new theory triumphs over an old one partly through propaganda (which stimulates scientists to make a "leap of faith") and partly through the passing of old-guard scientists.[29] In reality, however, he maintains that theories are compared by reference to common standards, but argues that these standards are rationally persuasive without being logically compelling. His critics have erred in equating absence of logical compulsion with absence of reason altogether. Nevertheless, Kuhn's standards are too general to offer explicit guidance to scientists faced with markedly contrasting theories. Criteria such as problem-solving ability and simplicity can be valued and interpreted very differently by different scientists. This makes Kuhnian theory-choice a more personal affair than Kuhn apparently realizes. More importantly, the standards are formal rather than substantive; Kuhn offers no standards for assessing the content of theories. As a result, scientists have no explicit criteria for deciding which theory advances a science in the direction it ought to take. Such a decision can be made only by referring to the intermediate or long-term goals of the science. I shall return to this important point in the next chapter.

SUMMARY

Because science seeks verified knowledge, it ought to be rational. Because science is collective, it cannot be rational unless it seeks collective agreement. But such agreement does not suffice for rationality, for men also can be collectively irrational. Many philosophers hold that, to be rational, science must aim at empirical growth, but they disagree as to how such growth should be achieved.

According to the logical empiricists, science grows cumulatively without overthrows: scientists generalize laws from statements of fact, explain laws by means of theories, and incorporate earlier theories within later ones, of which the earlier become special cases.

Popper replies that scientific growth always is revolutionary. Science advances rationally by putting forward bold, conjectural theories and then testing them as rigorously as possible, retaining only those that survive this process, and regarding them not as more probable but as not yet falsified.

Kuhn criticizes Popper for failing to recognize that much important science is not revolutionary at all. Science, he says, grows rationally through successive periods of normal science—in which a fundamental theory is extended—interrupted by revolutions, when that theory is replaced by an altogether different one bringing new problems for research to investigate. Scientists choose between the theories mainly according to standards that outlast revolutions but that different scientists value differently, so that the choice ultimately is made for good but differing reasons.

I have criticized these thinkers but will postpone my final appraisal until the close of the next chapter. In the meantime let us examine the reactions of philosophers to the clash between the Kuhnian and Popperian schemes.

NOTES

1. Rudolf Carnap, *Logical Foundations of Probability*, 2d ed. (Chicago: University of Chicago Press, 1962); Hans Reichenbach, *The Rise of Scientific Philosophy* (Berkeley and Los Angeles: University of California Press, 1956), ch. 16.

2. Ernest Nagel, *The Structure of Science*, pp. 336–37.

3. See Carl G. Hempel, "Studies in the Logic of Explanation," in *Readings in the Philosophy of Science*, ed. Herbert Feigl and May Brodbeck (New York: Appleton-Century-Crofts, 1953), p. 321. For a more recent statement, see Ernest Nagel, "Issues in the Logic of Reductive Explanation," in *Contemporary Philosophic Thought: Mind, Science, and History*, ed. Howard E. Kiefer and Milton K. Munitz (Albany: State University of New York Press, 1970), p. 121. I discuss reductionism more fully in ch. 6.

4. Carnap, *Logical Foundations of Probability*, p. 243: "For instance, we cannot expect to apply inductive logic to Einstein's general theory of relativity, to find a

numerical value for the degree of confirmation of this theory. . . . The same holds for the other steps in the revolutionary transformation of modern physics . . . an application of inductive logic in these cases is out of the question." See also *ibid.*, p. 571.

5. Popper's main works in the philosophy of science are *The Logic of Scientific Discovery* (originally published in 1936 as *Logik der Forschung*); *Conjectures and Refutations*; and *Objective Knowledge: An Evolutionary Approach* (Oxford: Clarendon, 1972). On Popper, see *The Philosophy of Karl Popper*, ed. Paul Arthur Schilpp, The Library of Living Philosophers, vol. 14 (La Salle, Ill.: Open Court, 1974). This is a collection of essays by various people and includes Popper's intellectual autobiography.

6. Popper, *Conjectures and Refutations*, p. 51.

7. *Ibid.*, p. 36.

8. Popper holds that a refuted theory may be retained until a better alternative is found. At one point he says that "if accepted basic [i.e., test] statements contradict a theory, then we take them as providing sufficient grounds for its falsification *only if they corroborate a falsifying hypothesis at the same time*" (*Logic of Scientific Discovery*, p. 87, emphasis added). Nevertheless, his general position is that a theory may be falsified in the absence of an alternative, and that as soon as it is falsified an alternative should be sought.

9. Popper, *Conjectures and Refutations*, pp. 36–37.

10. *Ibid.*, p. 220.

11. See Errol E. Harris, *Hypothesis and Perception: The Roots of Scientific Method* (New York: Humanities Press, 1970), pp. 75–76.

12. See Imre Lakatos, "Popper on Demarcation and Induction," in *Philosophy of Karl Popper*, ed. Schilpp, p. 247.

13. Kuhn's most important writings are *The Structure of Scientific Revolutions*; "Second Thoughts on Paradigms," in *The Structure of Scientific Theories*, ed. Frederick Suppe, pp. 459–82; "Logic of Discovery or Psychology of Research?" and "Reflections on My Critics," in *Criticism and the Growth of Knowledge*, ed. Imre Lakatos and Alan Musgrave, pp. 1–23 and 231–78.

14. Margaret Masterman notes 21 different senses in which Kuhn uses the term "paradigm" in the original edition of *Structure of Scientific Revolutions*. See her essay, "The Nature of a Paradigm," in *Criticism and the Growth of Knowledge*, ed. Lakatos and Musgrave, pp. 59–89.

15. Kuhn, "Postscript—1969," *Structure of Scientific Revolutions*, p. 190. By the mid nineteenth century, astronomers had observed that the perihelion (point nearest the sun) of the orbit of the planet Mercury advances 43 more seconds of arc per cen-

65

tury than can be accounted for by the perturbing effects of other planets. Einstein's general theory of relativity predicted this deviation exactly and explained that it follows from the fact that, being the planet closest to the sun, Mercury moves round it faster than the others under the greater influence of its gravitational field.

16. Kuhn, *Structure of Scientific Revolutions*, pp. 111, 121: "Paradigm changes do cause scientists to see the world of their research-engagement differently. . . . Though the world does not change with a change of paradigm, the scientist afterward works in a different world."

17. Kuhn, "Reflections on My Critics," pp. 257–58.

18. Kuhn, *Structure of Scientific Revolutions*, p. 166: "In its normal state . . . a scientific community is an immensely efficient instrument for solving the problems or puzzles that its paradigms define."

19. Kuhn, *ibid.*, pp. 118–25, 200.

20. *Ibid.*, p. 202: "Each may, that is, try to discover what the other would see and say when presented with a stimulus to which his own verbal response would be different."

21. *Ibid.*, pp. 155, 185, 199.

22. *Ibid.*, p. 204.

23. *Ibid.*, p. 159. I find this thesis persuasive. To my mind, intuitions are acts whose content is so condensed that its rational structure often is concealed. Instead of dismissing such acts as merely subjective, we should seek to reconstruct their underlying rationality. (See ch. 5.)

24. Some of the best criticism of Kuhn is to be found in *Criticism and the Growth of Knowledge*, ed. Lakatos and Musgrave; and in Dudley Shapere, "The Structure of Scientific Revolutions," *Philosophical Review* 73 (1964): 383–94, and "The Paradigm Concept," *Science* 172 (14 May 1971): 706–10.

25. Pulsars emit radiation in pulses of a few hundredths of a second at intervals of rather less than a second. They are thought to be rotating neutron stars, collapsed stars with a diameter of about 30 miles and a density roughly 1000 times that of the sun. They pulsate with astonishing regularity, and if used as clocks they would be accurate to within a fraction of a second per year.

26. See also Larry Laudan, *Progress and Its Problems*, pp. 75, 97–98.

27. Clifford M. Will, "Gravitation Theory," *Scientific American* 231 (November 1974): 25–33.

28. According to the neutralist theory (propounded by, e.g., J. Crow of the United States and M. Kimura of Japan), many mutations occurring in DNA are neutral rather than adaptive or maladaptive. Such mutations spread through populations in a process called "genetic drift" and are responsible for many traits.

29. See the criticisms made by Popper, Watkins, and Lakatos in *Criticism and the Growth of Knowledge*, ed. Lakatos and Musgrave; also Imre Lakatos, "History of Science and its Rational Reconstructions," in *PSA 1970: In Honor of Rudolf Carnap*, Boston Studies in the Philosophy of Science, vol. 8, ed. Roger C. Buck and Robert S. Cohen (Dordrecht, Holland: Reidel, 1971); and Israel Scheffler, *Science and Subjectivity* (Indianapolis: Bobbs-Merrill, 1965), p. 18. Kuhn replies to these and other criticisms in the "Postscript" to his *Structure of Scientific Revolutions* and in the other works of his that I have cited.

Chapter 4

Research Programs to Metaphysical Blueprints

The Popper-Kuhn debate confronted philosophers of science with the following problem. Popper maintained that science is rational to the extent that it seeks to criticize its theories. Kuhn, on the other hand, argued that for most of its lifetime a major theory is, and should be, developed rather than criticized. Could these views be reconciled? Two of Popper's former students, Imre Lakatos and Paul Feyerabend, decided they could. But they proposed quite different solutions: self-styled rationalism on the one hand, conscious anarchism on the other. I begin with Lakatos.[1]

IMRE LAKATOS

Popper is right, says Lakatos, to oppose the imperialism of single, uncriticized theories, but wrong to advocate ruthless refutation. Kuhn is right to exempt a developing theory from criticism but wrong to exempt an entire field. The real test of a theory is this: does it successfully predict new facts? If it does, "refutations" can be ignored. Science is kept critical through competition between theories. Rival theories expose one another's weaknesses by their own successes, the less fertile theories eventually being abandoned.

Lakatos calls a developing theory a "research program"; it consists of a "hard core," a "protective belt," and a "heuristic." The hard core comprises the assumptions of the program—in Newton's case, the three laws of motion and the law of gravitation. The protective belt is a collection of auxiliary hypotheses which keep the hard core unrefuted. Instead of dropping an axiom from the core, the scientist adds or sheds an auxiliary hypothesis. The Newtonian protective belt included geometrical optics, Newton's theory of atmospheric refractions, and many other hypotheses. A heuristic is a research policy indicating how the implications of the hard core may be deduced and applied to actual situations. It suggests what types of hypothesis to propose, what problems to solve, and what techniques to use in solving them. In following the heuristic, the scientist ignores all the anomalies except those the heuristic anticipates. In Lakatos's words,

"The scientist lists anomalies, but, as long as his research program sustains its momentum, ignores them. *It is primarily the positive heuristic of his program, not the anomalies, which dictate the choice of his problems.* Only when the driving force of the positive heuristic weakens, may more attention be given to anomalies."[2]

Newton's positive heuristic, says Lakatos, included (a) the principle that a planet is a gravitating spinning-top of roughly spherical shape and (b) a mathematical apparatus, involving the differential calculus, the theory of convergence, and differential and integral equations.[3] Guided by this heuristic, Newton worked out a series of increasingly complex models for calculating planetary orbits. Treating the sun and planets as mass-points, he constructed first a planetary system of a sun and a single planet, then a system in which both sun and planet revolved about a common center of gravity, then a system with more than one planet but with only heliocentric and not interplanetary forces. Next he switched from mass-points to mass-balls—a difficult mathematical move that delayed publication of the *Principia* by more than a decade. After solving this problem, he turned to spinning balls and their wobbles. Introducing interplanetary forces, he calculated perturbations, then postulated bulging planets rather than round ones, and so on. All these models entered as hypotheses into the protective belt.

For Lakatos, a research program either progresses or degenerates. It progresses if each change in the protective belt leads to some new and successful prediction. It degenerates if it ceases to make and confirm unexpected predictions and instead accounts for new facts with ad hoc hypotheses unanticipated in its heuristic. (An ad hoc hypothesis predicts only those facts it has been invented to explain;[4]

hence it does not promote scientific growth.[5]) Between 1913 and 1921 the Bohr-Sommerfeld program, investigating the structure of the hydrogen atom, successfully predicted one fact after another, including the Rydberg constant (for atomic spectra similar to the hydrogen spectrum), the Balmer series (of spectral lines of hydrogen), the Pickering-Fowler series, the Stark effect (the splitting of spectral lines of hydrogen in an electric field), and the normal Zeeman effect (the splitting of such lines in a magnetic field).[6] In 1922 the program began to degenerate. Bohr's formula for the spectra of diatomic molecules was refuted and replaced by a formula that was correct but ad hoc. Then unexpected double lines turned up in the alkali spectra. These were accounted for by an ad hoc "relativistic splitting rule" and then by an electron spin that was inconsistent with special relativity. Bohr's program was reacting, unsuccessfully, to new discoveries instead of predicting them. It soon was overtaken by the wave mechanics of de Broglie and Schrödinger and by Heisenberg's matrix mechanics.[7]

A research program is judged by its performance in comparison with rivals. One program overtakes another, says Lakatos, if it successfully predicts all that its rival correctly predicts and more besides. Einstein proposed his special theory of relativity in 1905, but his program did not supersede Lorentz's ether program until 1915, when Einstein successfully explained the anomalous precession of Mercury's perihelion.[8] The Newtonian and Huyghensian programs in optics battled for nearly a century and a half before a majority of physicists were persuaded to the latter by Fresnel's experiments.[9]

However, any program can make a comeback, and one can never know at the time whether a particular program is finished. For nearly a century Prout's program failed to persuade most chemists until it was vindicated by Rutherford and Soddy.[10] Only when a program is almost completely abandoned can one pick out the decisive test by which it was superseded. According to Lakatos, the crucial nature of Young's double-slit experiment was not recognized until a half generation later.[11] There is, to repeat, no "instant rationality," no principle by which one can tell at the time whether a given program should be abandoned.

CRITIQUE OF LAKATOS

As a research tool, Lakatos's theory has no superior. Lakatos directs the researcher to look for specific items in a scientific tradition: the hard core, the protective belt, the heuristic, the progressive and degenerating phases. Moreover, he is always lively and readable. As a philosopher, however, he is somewhat narrow. He does not examine the evolution of science as a whole, and he has little feeling for the aesthetic side of science, which has stimulated other thinkers. He spends too much time debating with opponents both real and imagined. His early death deprived him of the chance to transcend these limitations.

Lakatos's theory has specific weaknesses. In treating research programs as competitors, he does not explain how a major research tradition, such as the Newtonian tradition in mechanics, can guide a discipline or field for decades without meeting a serious rival. He also fails to account for revolutions. For him, one research program simply overhauls another after a more or less lengthy struggle. There is no crisis precipitating a search for alternatives, and no awareness, when a majority switch to a new theory, that a revolution has taken place. Yet, as Kuhn has shown, such crises have occurred, as in the search for a quantum theory to replace that of Bohr.

Although Lakatos frequently judges past science, he offers present scientists no consistent standards to follow. He distinguishes between progress and degeneration in a research program, but then says that a degenerating program can hit back at any time, and then again insists that final defeat can only be identified long after the event. The last two propositions mean that there is no way to tell whether a certain degenerating program will recover. Equally, then, there is no means of judging whether a currently progressive program will not eventually be defeated by a program now in the doldrums. Lakatos tells the scientist how to recognize progress and degeneration but not what to do about them. He suggests, it is true, that editors should reject papers submitted by workers in degenerating programs, and that funds should be denied such programs.[12] Yet here he is inconsistent. Having declared that only hindsight will reveal when a

71

given program should have been abandoned, he advises the scientific establishment to do just that which may kill a program, deservedly or not. What if a rejected paper is one that would have begun an upward trend?

Some will say that it is enough to provide criteria for appraising programs without offering advice on how to do "good" science. Yet criteria of appraisal cannot avoid being oblique directives for practice. To say that a program is progressive or degenerating is to imply that it deserves more support, or less, from scientists. Lakatos in effect cancels this implicit directive by his provision for a comeback. Hence he should have reserved for historians the terms "progressive" and "degenerating," and for contemporaries "temporarily progressive" and "temporarily degenerating," expressions whose appraisive force is minimal.

Lakatos also fails to explain why a new research program is started at all. For Feyerabend, as we shall see, scientists put forward new theories to predict facts that will refute an established theory. For Kuhn, they propose them when they lose confidence in that theory. But for Lakatos a new research program comes into being in response to no particular stimulus, not even the degeneration of an existing program.

More importantly, as Maxwell points out, Lakatos proposes no way of inventing a hard core and a heuristic rationally or of choosing rationally between hard cores and heuristics. This makes scientific growth strikingly nonrational. As I have said, to act rationally is to form a defensible aim and to act so as to attain it. Yet Lakatos nowhere proposes that a science or a specialty might have an aim and that research programs might be launched to realize it. For him, there is no way to judge, before testing, whether a proposed program is likely to be empirically successful, because there is no publicly agreed or publicly discussible aim in relation to which the program can be appraised. Picking a promising program depends, it seems, on luck or guesswork, not on rational discussion.

Lakatos's omission is the more striking in that his comeback provision makes it impossible for contemporaries to judge a research program by its empirical record alone. Elie Zahar, Lakatos's pupil, makes the crucial point that what led brilliant physicists and mathematicians like Max Planck and Hermann Minkowski to switch from Lorentz's program to Einstein's before the latter became empirically progressive was the greater power of Einstein's heuristic.[13] Whereas Lorentz stipulated that electromagnetic and molecular forces obey his transformation equations, Einstein required that all forces obey these equations and that any new law imply a classical law as a limiting case. Thus Einstein's program promised to predict more facts than Lorentz's. Yet Zahar misses the chance to make a "creative shift"[14] in Lakatos's own research program. Instead of arguing that all programs should be appraised for both intellectual coherence and empirical success, Zahar makes the point for this case only.

PAUL K. FEYERABEND

Calling himself an "epistemological anarchist," Paul Feyerabend is the most polemical and provocative of the writers we are considering. His intention, he announces, is "to convince the reader that *all methodologies, even the most obvious ones, have their limits,*"[15] and that the scientist should be free to try any procedure he likes. In theoretical arguments, and in his case study of Galileo, he seeks to show how science stagnates under the domination of single theories, and he proposes ways to overthrow such theories.

Once a comprehensive theory (e.g., Aristotle's, Ptolemy's, Newton's) is widely accepted, he argues, it encourages scientists to propose theories that cohere with it and yield predictions consistent with its own. Feyerabend calls these further theories and the fields of research they govern "auxiliary sciences." The longer a theory is accepted, the more it and its auxiliary sciences reinforce one another. What does this imply for a new opposing theory?

A new theory is tested, says Feyerabend, not only against data gathered to verify its own predictions but against data provided by auxiliary sciences based on assumptions similar to those of the established theory. The Copernican theory, for instance, was confronted by data supplied by auxiliary sciences dealing with the process of perception, the earth's atmosphere (in which the tests were made), the properties of the earth (from which the tests were conducted), and so forth.[16] These data reflected assumptions similar to those of the Ptolemaic theory which the Copernican theory was designed to refute.[17] For example, the Copernican claim that the earth rotates on its axis, a motion that cannot be detected by an earthbound observer since he himself shares in it, inevitably was "refuted" by the data reflecting the Aristotelian assumption that all motion can be observed. How, then, can an established theory be overthrown?

Feyerabend answers this question in two ways. In principle, he says, the scientist may do whatever he likes. There is no rule of research that has not been broken at some time in the best interests of science; hence one cannot insist that in a given situation the scientist must follow a certain course. After all, this may be just the situation in which the rule should be broken. Such is the essence of Feyerabend's well-known dictum, "Anything goes." In his own words: "There is not a single rule, however plausible, and however firmly grounded in epistemology, that is not violated at some time or other. Such violations are not accidental events. . . . On the contrary . . . they are necessary for progress. . . . There is only *one* principle that can be defended under all circumstances, and in *all* stages of human development. It is the principle: *anything goes.*"[18]

Nevertheless, he argues, it is often reasonable to proceed "counterinductively"—to invent a theory that outflanks the assumptions permeating the known facts by predicting a different set of facts plus those known facts that the established theory cannot explain. Where are new facts to be found?

The new theory will predict fresh facts, some of which may be confirmed, through not necessarily at once. The theory also will be accompanied by subordinate hypotheses, which may grow in time into auxiliary sciences able to predict further facts. Thus Galileo, for instance, promoted the Copernican theory by proposing new astronomical data, gathered through his telescope, to bypass the existing data supporting the Ptolemaic theory, as well as a new dynamics (that natural motion is circular and shared motion unobservable) to replace the Aristotelian dynamics likewise supporting the Ptolemaic theory. But the Copernican theory also needed further auxiliary sciences, such as meteorology and physiological optics (to provide a theory of telescopic vision).[19] These developed more slowly. According to Feyerabend, each successful prediction yielded by the new theory encourages scientists to work further with the theory's incipient auxiliary sciences, and vice versa.

How, then, does the innovating scientist persuade other scientists to take the new facts seriously? He uses ad hoc hypotheses and, believe it or not, propaganda. Ad hoc hypotheses explain provisionally facts that otherwise are unaccounted for. The scientist hopes that these hypotheses eventually will be confirmed and so provide the theoretical core of the auxiliary sciences he needs.

Galileo propped up the Copernican theory with several ad hoc hypotheses, notably the hypothesis that the telescope informs us about distant objects more accurately than the naked eye, the hypothesis that only relative motion is observed, and the hypothesis that the earth's motion causes the tides. Galileo used the first hypothesis to explain a number of observations he had made with his telescope, such as the fact that Mars and Venus look very much larger when approaching the earth. According to naked-eye observations, supporting the Ptolemaic theory, these planets dilate only slightly. Galileo claimed, however, that the telescope eliminates the irradiating rays that are caused by the planet's proximity to the sun and that seem to the naked eye to be part of the planet itself. These rays make the planet, when observed by the naked eye, appear larger than it really is when far from the earth. Thus, according to Galileo, the discrepancy between the (to the naked eye) slight enlargement of the planet as it approaches the earth and the much greater enlargement predicted for it by the Copernican theory actually results from the illusory magnification of the planet by the naked eye when the planet is close to the sun. Galileo's observations, then, conflicted with the established facts, with the Ptolemaic and Aristotelian theories (which assumed that the senses are trustworthy), and with common sense (which treated the Ptolemaic-Aristotelian assumption as a plain fact of life). To make his facts persuasive, Galileo argued that they were explained by his telescopic hypothesis and that they favored the Copernican theory. In this way, says Feyerabend, Galileo used two otherwise unsupported hypotheses—the Copernican theory and the telescopic hy-

75

pothesis—to support each other. Both hypotheses were ad hoc, but the fact that each reinforced the other made both more plausible than they would have been separately.

Galileo's hypothesis of relative motion also was ad hoc. Unlike the telescopic hypothesis, however, it was not proposed to account for any new facts. How did Galileo seek to make it persuasive? Feyerabend says he used propaganda. It long had been known that in a few cases shared motion is unobservable. For example, if two ships leave port at the same speed, an observer on one ship cannot see the other move, but can only see the shoreline slip behind. These cases had been regarded as exceptions to the general law that all motion is observable. To make his hypothesis persuasive, Galileo had to claim that the exceptions were really the rule. He did so, says Feyerabend, by means of "psychological tricks." In the dialogue justifying the principle of relative motion, he made one speaker use clever rhetoric to persuade the other of a palpable untruth, namely, that people once applied the principle to all cases of motion but have simply forgotten that they did so, just as (according to Plato) the Athenian slave boy had forgotten the knowledge that Socrates elicited from him. Although the principle of relative motion is true, says Feyerabend, Galileo justifies it neither by new arguments nor by new facts but solely by the false claim, made persuasive by rhetorical devices and the appeal to the authority of Plato, that we observe the principle without realizing it.[20]

Feyerabend also urges scientists to "proliferate theories." What science needs most, he says, is an increasing number of new theories, especially comprehensive revolutionary theories providing both a new view and (as we shall see) a new "experience" of a wide range of phenomena. Old theories are overthrown by refuting facts. To find such facts quickly, we must propose and test new theories that predict them. Consider the refutation of the second law of thermodynamics by the phenomenon of Brownian motion[21] (the random agitation of tiny particles suspended in a fluid in thermal equilibrium).[22] This phenomenon had been a long-standing anomaly to thermodynamics, which predicts that such particles will not create any osmotic pressure. Nevertheless, the refutation was not demonstrated until Einstein, in 1905, deduced the existence and magnitude of Brownian motion from the theory of statistical mechanics, and Jean Perrin, in 1910, confirmed this experimentally. In sum, where Kuhn would allow a theory to uncover the anomalies that eventually will bring it down, Feyerabend argues that normal science is too slow. To multiply the number of refuting facts, we must proceed counterinductively against an established theory with as many new theories as we can.

Feyerabend maintains that a new theory reclassifies old facts so that they reflect its assumptions. When a highly comprehensive theory (a "universal" theory or "cosmology")[23] reclassifies facts, it also recreates experience." When Galileo reclassified all visible motion as relative motion, he led those scientists who accepted his view, and eventually the educated public, to experience such motion as relative to the earth's motion rather than as part of it. Thus, says Feyerabend, Galileo's theory changed human experience by leading people to observe a phenomenon, in this case visible motion, differently. Therefore, concludes Feyerabend, there is no "stable and unchanging experience" common to all men throughout history.[24]

CRITIQUE OF FEYERABEND

Feyerabend states that he is out to demolish the law-and-order view of science. He denies that science is rational, and he offers no theory of scientific growth. Nevertheless, as part of his liberationist creed for science he has advanced a number of theses that are both explicit and partly true: for example, that a comprehensive theory reclassifies facts and recreates experience; that a new comprehensive theory needs new data and so new auxiliary sciences; that it is reasonable to proceed counterinductively; that a new theory needs a breathing space without criticism; and that theories should be proliferated.

In his advocacy of the principle "anything goes," however, Feyerabend is inconsistent. As he explains it, the principle is reasonable enough: no single rule is so important that it can supersede all others in all situations. He also maintains, again reasonably enough, that there are circumstances when it is "advisable" to proceed counterinductively, to promote a theory against the weight of the evidence, to proliferate theories, and to use propaganda and ad hoc hypotheses.[25] But beyond his study of Galileo he does not specify what these circumstances are, and Galileo's case is too exceptional to justify the sweeping lessons that Feyerabend draws from it. Moreover, Galileo was no anarchist. Instead of improvising, Galileo suspended one rule, "Turn anomalies to your theory into confirming instances," in favor of another, "Ignore at least some anomalies if you think your theory is true."

Feyerabend says in effect that the scientist may follow "counterrules" or not as he pleases. Although proliferation of theories is "the only method that is compatible with a humanitarian outlook," the scientist need adopt the rule only if he wishes to. Thus, says Feyerabend, "A scientist who wishes to maximize the empirical content of the views he holds and who wants to understand them as clearly as he possibly can must therefore introduce other values; that is, he must adopt a *pluralist* methodology." More strongly still, he declares that the "epistemological anarchist" has no firm commitments whatever except his opposition to universal standards: "His aims remain stable, or change as a result of argument, or of boredom, or of a conversion experience, or to impress a mistress, and so on. . . . His favorite pastime is to confuse rationalists by inventing compelling reasons for unreasonable doctrines." Indeed, Feyerabend makes it plain that his overriding aim is to "mock rationalists." He advises his readers always to remember that "the demonstrations and rhetoric used do not express any 'deep convictions' of mine. They merely show how easy it is to lead people by the nose in a rational way."[26]

This is "anything goes" with a vengeance, for with this remark Feyerabend undercuts the reasonable position he had taken earlier. If the anarchist scientist is entitled to do as he pleases, why should it ever be "advisable" for him to proceed counterinductively? If he wishes to "impress his mistress" by sticking to an established theory or counterattacking Feyerabend, why should he not do so?

Apparently, Feyerabend wants to have it both ways. He advocates complete freedom for the scientist, but he also recommends certain "counterrules" as worth following. The contradiction between these proposals is hidden, because Feyerabend does not specify the situations in which the counterrules might apply. Again, in submitting rules of any kind, Feyerabend implies that it is *not* reasonable for the scientist *always* to do whatever he likes. But he also implies that the scientist can use these principles as he pleases and therefore that it *is* reasonable for him to do whatever he likes.

What should Feyerabend have concluded? To be consistent, he should have concluded that different rules apply in different situations and that no rule applies universally. He should have maintained that all particular rules remain optional, that any particular rule usually is worth following in a certain type of situation without being mandatory for it, and that it is up to the individual scientist to decide which rules are most suited to his situation.

Again, Feyerabend presses his case for proceeding counterinductively and proliferating theories to the exclusion of other possibilities.[27] Something like Kuhnian normal science sometimes occurs and is sometimes justified. But Feyerabend neither explains normal science nor justifies it under any circumstances. Yet it was surely reasonable for the majority of scientists to pursue research in mechanics under the Newtonian tradition until the mid nineteenth century. It also would have been reasonable for at least some scientists to have worked on alternative theories in mechanics during the Newtonian hegemony. If they had done so, and if one of these theories had anticipated Einstein's relativity, Newton's theory would have been superseded sooner. What Feyerabend does not admit is that (1) a comprehensive theory may be a good one and may rightly guide the bulk of research in a field for a considerable time, and (2) such a situation is quite compatible with at least some proliferation of alternative theories.

Theory-proliferation is admirable in fields such as particle physics that are brimming over with facts but lack a successful theory to explain them, and in fields containing serious anomalies. But often it may simply be too expensive. Grant committees must decide which theory to back when theory A is technically so expensive that theories B and C must be abandoned if A is chosen, or when theory A predicts many new facts but is expensive to test, while theory B requires little apparatus but is otherwise much less promising. Feyerabend nowhere considers the financial constraints on research.

Finally, for good or ill, scientific theories alter our "experience" less than Feyerabend thinks. In ordinary life we do not perceive the motion of objects as the relative part of a more inclusive motion, though we may do so occasionally. Again, notwithstanding quantum mechanics, we still experience the world as populated by substantial objects with stable properties. The people whose experience is changed by a theory are mostly the scientists who work with it.

NICHOLAS MAXWELL

The problem facing Lakatos and Feyerabend was how to reconcile the best insights of Popper and Kuhn. But the solutions they reached have posed further problems: in Lakatos's case, how to make a rational choice of hard cores, in Feyerabend's, how to limit theory-proliferation. These problems are at the heart of the present debate on scientific change and rationality. I turn first to the choice of hard cores.

For Lakatos, as we have seen, a research program can be appraised rationally once it has been defeated but cannot be chosen rationally in advance. Consequently, contrary to Lakatos's belief, scientific growth becomes strikingly irrational, since it is initiated by guesswork, not rational discussion. For Kuhn, the first supporters of a new theory are moved by faith rather than by reasons that can be discussed publicly. Feyerabend, finally, welcomes the irrationality of science, although he makes some provision for theory-comparison.[28] For all these writers, there are some rational methods of theory-appraisal, but there is no rational method of theory-invention. Can such a method be found?

Indeed it can, says Nicholas Maxwell; it is the method of "aim-oriented empiricism."[29] Maxwell begins by criticizing the philosophy of "standard empiricism" which, he claims, underlies previous theories of scientific change. According to standard empiricism, the aim of science is to discover more and more about the world without making metaphysical (i.e., untestable) assumptions about what the world is really like. The only evidence that a theory is true is empirical confirmation, and there is no way of conjecturing beforehand whether any theory is likely to be more true than another. All we can

80

do is launch research programs and see which of them successfully predicts the most new facts (i.e., makes novel predictions that are confirmed).

Yet, argues Maxwell, standard empiricism is refuted by scientific practice, for scientists regularly reject (or seek to replace) aberrant, ad hoc, inconsistent theories even when they are confirmed. (Consider the long history of controversy over the foundations of quantum mechanics.) But one cannot prefer simple, harmonious, consistent theories even when the facts are against them, unless we assume that at a deeper level nature itself is simple and harmonious. For example, instead of accepting that matter consists of some 200 different types of elementary particles, physicists are looking for a much simpler underlying pattern. Yet scientists continue to pay lip service to standard empiricism, and refrain from publicly discussing the merits of embryo research programs before they are tested. Saddled with this methodology, science is less rational and less empirically successful than it might be. What is the alternative?

Science, declares Maxwell, proceeds in practice from the metaphysical assumption that the universe is intelligible. From this metaphysical "blueprint" other, more specific, blueprints are derived, as scientists seek to find out what this intelligibility is like. Scientists therefore need to propose, appraise, and choose among different possible blueprints, both for science as a whole and for particular sciences. Each blueprint is a proposal to develop theories to articulate a certain kind of order presumed to exist in the world. Here are some blueprints that might be considered: that the universe is activated by a single being, God; that the world is informed by a cosmic purpose; that the material world is an imperfect copy of an ideal one (Plato); that the world is full of things striving to realize their potentials (Aristotle); that nature is composed of a few fundamental types of physical entity interacting in accord with a few fairly simple mathematical laws (Galileo); that the universe is made up of infinitely hard corpuscles interacting by contact (Descartes); that the universe is composed of mass-points that interact via central forces of attraction and repulsion varying with distance (Newton, Boscovich, Helmholtz); that nature consists of a single unified field (Einstein).[30]

How are we to choose among rival blueprints before we have

tested any theories that can be developed from them? If we can devise a rational *a priori* procedure for choosing among blueprints, says Maxwell, we shall have the basis of a rational method for inventing theories and hence a rational method of discovery.

According to Maxwell, we should choose the simplest, most intelligible blueprint and then make it precise and testable by constructing theories that develop its implications and meet with increasing empirical success. If our blueprint reflects at least roughly the way we think things are, it is likely to yield theories that successfully predict more and more phenomena. If it does not, we soon shall cease to turn up new facts. Contrast the stagnation of science under the Aristotelian and Platonic-Ptolemaic blueprints with its progress under Kepler, Galileo, and Newton, who, says Maxwell, chose blueprints with some correspondence to the way things seem to be. To ensure that we have the best blueprint for this science at this time, we must choose our blueprint with care and appraise it continually.

A mature science, says Maxwell, should improve its chosen blueprint through continuous, critical discussion. In particular, scientists should strive to develop new blueprints *before* they are needed, so that revolutionary new theories are available when current theories break down. Unfortunately, few scientists have done this. Those who have include Faraday with his unified field and Einstein with his two principles for future theories to include—relativity and general covariance. John Archibald Wheeler, working from general relativity, has proposed the blueprint idea that matter is a manifestation of the curvature of space.[31] Roger Penrose, seeking to unify general relativity and quantum mechanics, has suggested that points in space-time arise from the intersection of the paths of massless particles.[32] Most scientists, however, do not even discuss blueprints, on the assumption that they are speculative, mystical, or obscurantist.

How are we to choose the best blueprint? Maxwell proposes six rules:[33] (1) Other things being equal, choose the simplest, most coherent blueprint. (2) If this blueprint is empirically successful (a sign that it represents the way things are), narrow it, defining more sharply the type of theory to be proposed and tested, even perhaps making the blueprint itself a testable theory. (3) If the blueprint has little or no empirical success, broaden it, increasing the range of permissible theories in the hope that at least one of them will prove successful. (4) Avoid outlandish blueprints whose offspring theories seem likely to be exceedingly difficult to test. (5) Choose rules of theory-acceptance most likely to lead to the successful articulation of the best blueprint (e.g., in physics, accept action-by-contact theories, reject action-at-a-distance theories, accept mechanistic theories, reject teleological ones).

These five rules imply that blueprint discussion will center on three main types of problems: empirical problems, created by anomalous facts; theoretical problems, arising from conflicts between accepted theories or from incompatibilities between those theories and the blueprint itself; aim-improvement problems, caused by the discovery of arbitrary, incoherent, or inconsistent features within the blueprint. Hence there arises a further rule: (6) Construct new theories by locating and solving problems of these kinds, especially aim-improvement problems. Paul Dirac, for instance, proposed his relativistic quantum theory in an attempt to reconcile special relativity with quantum mechanics.

How does a blueprint yield testable theories? A blueprint provides rules of theory-acceptance—such as the conservation laws and symmetry principles of modern physics—that specify what future theories should be like (e.g., what form of simplicity they should take).[34] The blueprint also supplies a special terminology, often mathematical, which defines precisely the unchanging properties of the entities the blueprint postulates. The scientist's task is to construct a theory that attributes a simple pattern to phenomena and takes on a simple form when expressed in this terminology. For example, the field equations in Einstein's general relativity follow uniquely from his rules of theory-acceptance and the rules of the tensor calculus that he used as a language. If they had not been expressed in the tensor calculus, these equations would have been almost unintelligible.[35]

Owing to the neglect of blueprint study, says Maxwell, physics today is in disarray with no unified aim. The blueprint ideas behind general relativity and quantum mechanics are incompatible. General relativity is microrealistic and field theoretic; quantum mechanics is indeterminist. Quantum mechanics will not be consistent until it has been made microrealistic, a theory about microsystems interacting with microsystems. At the moment we have no idea what a nondeterminist, microrealistic theory would be like. The foremost aim-improvement task of physics today is to develop general ideas for such a theory, and then for a theory that is field theoretic in addition.[36]

CRITIQUE OF MAXWELL

Maxwell's theory of aim-oriented empiricism is the outstanding work on scientific change since Lakatos, and his thesis is surely correct. Scientific growth should be rationally directed through the discussion, choice, and modification of aim-incorporating blueprints rather than left to haphazard competition among research traditions seeking empirical success alone.

To understand Maxwell's advance, contrast him with others on the issue of challenging an accepted theory. Kuhn, Feyerabend, and Lakatos all agree that the way to refute the assumptions of an established theory is to provide facts falsifying the predictions of that theory. But if the assumptions are the real target, why not attack them directly? As Maxwell points out, empirical refutations only tell us that there may be flaws in the assumptions. We should therefore examine the assumptions, and if we find them unsatisfactory, we should construct more comprehensive and coherent ones and launch a research program to test them.

Nevertheless, in its present form Maxwell's theory has certain weaknesses. The key terms "intelligibility," "beauty," "simplicity," and "coherence" are used interchangeably. Yet however beautiful a simple theory may be, its beauty and simplicity are not identical. Similarly, if a theory is simple, it presumably is coherent and intelligible. But the three properties are distinct. Admittedly, Maxwell proposes criteria for blueprint intelligibility—namely adequacy, objectivity, nonarbitrariness, generalization, and coherence[37]—but these criteria only emphasize the need to distinguish between a blueprint's intelligibility and its other properties.

Maxwell fails to stress sufficiently that competing aims can be compared only in relation to an overriding aim. He provides such an aim for physics in the quest for the ultimate invariant constituents of matter and their invariant properties (or for nature's fundamental pattern), but he does so only in passing. The point to drive home is that, according to his theory, every choice between blueprints for physics is specifically a choice between proposals of what constituents and properties to look for.

Maxwell also does not consider seriously enough the possibility that science might have some permanent rules of theory-acceptance. Gary Gutting, for example, has proposed that there is an "untouch able core" of methodological rules, which he calls the "founding intentions" of science, or the intentions science must fulfill if it is to provide a rational understanding of the world.[38] These rules state that a scientific theory should be publicly discussible, empirically testable (at least in principle), logically consistent, tolerably precise, fairly comprehensive, and so forth. When a conflict arises between two fundamental research traditions with different specific rules, a choice must be made by considering which tradition, given present re sources, is likely to fulfill these intentions best.[39]

For example, says Gutting, the traditions of Galileo and Newton rejected the Aristotelian demands for teleological explanations and consistency with common sense. Similarly, the quantum tradition broke with the classical requirement of providing a determinate explanation of each observable event. Yet Galileo and Newton could reasonably claim that new methodological rules calling for mathematical laws and idealizations were more likely than the old rules to fulfill the founding intentions directing scientists to propose theories that are precise and comprehensive. For instance, the new rules were likely to lead to precise descriptions and predictions of phenomena, such as falling bodies and projectiles, of which the Aristotelian account was vague and inaccurate. They also were likely to produce precise accounts of phenomena, such as comets and certain optical effects, which the Aristotelians had been unable to explain at all. Similarly, many quantum scientists could claim, and did, that at present only a nondeterminist theory would give a mathematically precise account of atomic and subatomic phenomena. Until critics

85

produce a determinist alternative, they said, we must either use quantum mechanics or abandon the attempt to provide detailed mathematical explanations of such phenomena.

Finally, Maxwell pays little attention to science at the level of the research tradition, though he may correct this omission.[40] At this level there arises the problem posed by Feyerabend's work—how to limit theory-proliferation. We need to find a middle way between domination by a single tradition, recommended by Kuhn, and unlimited multiplication of traditions, permitted by Lakatos and advocated by Feyerabend. Does Maxwell's theory have anything to say to this problem?

At first glance Maxwell might seem to support single-theory rule Kuhnian style. First choose your blueprint; then choose the theory most likely to realize it; if the theory turns out to be clumsy or empirically unsuccessful, change it; but don't allow competition between theories. Indeed he says:

> Science should be seen as one gigantic research program which takes as its aim the articulation of a presupposed metaphysical system chosen for its inherent simplicity and intelligibility. Viewing science as this kind of monolithic research program violates the crucial Popper-Lakatos requirement that there must be *competing* theories or research programs if there is to be critical, rational progress towards the truth.[41]

Nevertheless, Maxwell's theory is compatible with more competition than he indicates, though it is a weakness of the theory in its present form that this implication is undeveloped. Let me explain this point.

For Maxwell, the important competition takes place at the blueprint level. If a science is to be ready at all times to modify or replace its current blueprint, new blueprints continually must be proposed. For a blueprint can be appraised only in relation to other possible blueprints, including the overriding blueprint of the science. At this level, I suggest, unlimited proliferation is reasonable, for blueprints do not need to be tested empirically, and it is only the shortage of funds for empirical testing that makes it desirable to limit the proliferation of research traditions.

Does Maxwell's theory permit competition at the level of the research tradition? Maxwell maintains that an empirically unsuccessful blueprint may need to be broadened, increasing the variety of types of theory that can be considered for testing. At this point, though Maxwell does not say so, competing theories can be submitted for consideration, and if they are found equally intelliglble, can be tested simultaneously. Thus an opportunity is provided for competition between research traditions.

GARY GUTTING

How in practice might a blueprint guide a research tradition? Gary Gutting has shown how the Newtonian tradition—in Maxwell's terms, the tradition working from the Newtonian blueprint—fostered the construction in various fields of a range of theories, or subordinate traditions, all logically independent of the major theory yet consistent with it (even when inconsistent with one another, like thermodynamics and statistical mechanics).[42] All these theories included Newton's laws of motion; used the Newtonian concepts of position, time, energy, and so on; shared a common macroscopic observation language; and observed the same methodological rules. Nevertheless, each theory developed and interpreted the Newtonian core uniquely. When Max Planck, for instance, sought to solve the problem of black-body radiation, he could draw on several theories—electromagnetic theory, thermodynamics, statistical mechanics—all consistent with the Newtonian theory, yet capable of being combined in different ways.

According to Gutting, the Newtonian tradition shows us how to steer a midcourse between single-theory dictatorship and the anarchistic proliferation of rival theories. The solution, he says, is to find an overall "conceptual framework" (hard core, blueprint) broad enough to promote a range of subordinate theories.

Such a framework operates as a general schema capable of being specified by many different (sometimes mutually inconsistent) theories. It is "not so much a *description* of the world as a *programme* for developing such a description."[43] The framework principles are not so much laws describing the behavior of things as rules prescribing the form such laws must take. In the Newtonian framework these principles include the law of motion, $F = ma$, and the principle of energy conservation. In these principles the central concepts of force and energy are specified only to the extent that each must form part of a single quantitative relationship. Thus it is possible for subordinate theories to define many different kinds of force and energy.

A framework can be developed by "extension" or by "synthesis." It is extended by proposing and deploying subordinate theories, consistent with the framework, for various domains within the scope of the framework. Thus the Newtonian framework led to different theories for planetary motion, hydrodynamics, acoustics, and so forth. Although Gutting does not say so, extension may also include intertheory competition within a domain, as in the rivalry between the British field tradition in electromagnetism and the Continental action-at-a-distance tradition, each of which incorporated different elements from the Newtonian framework.

Synthesis, for Gutting, is the attempt to combine two or more subordinate theories within a more comprehensive theory. Some examples are Newton's original union of Galileo's laws of terrestrial motion with Kepler's laws of celestial motion, the unification of thermodynamics and mechanics in statistical mechanics, Planck's attempt to combine electromagnetic theory with statistical mechanics, and Einstein's attempt to unite electromagnetic theory with mechanics. The syntheses of Newton, Planck, and Einstein produced scientific revolutions. Why?

Gutting states that whereas extension only applies the framework principles further, synthesis requires the scientist to combine two highly specific theories, a problem to which there are few solutions, and which therefore may lead to the emergence of a radically different theory.[44] In short, whereas extension only requires the scientist to apply the concepts and principles of the framework, synthesis may compel him to examine them.

Gutting shows us, then, how Maxwell's blueprint might be applied at the level of empirical investigation. Nevertheless, his theory has a serious weakness—his conceptual framework is protected from external criticism. Synthesis is only indirectly a critical mechanism. So long as synthesis succeeds within the framework, the flaws of the framework remain concealed. A stronger form of framework criticism is needed, and this is what Maxwell provides. Gutting's framework is not unlike a Maxwellian blueprint. His framework principles resemble blueprint principles which specify the forms of acceptable theories. But a Maxwellian blueprint is chosen only after rigorous comparison with rivals. In Gutting's theory, framework weaknesses eventually are revealed by synthesis; in Maxwell's they are the direct object of continuous criticism. Thus in Gutting a revolutionary new theory still comes as a surprise, whereas in Maxwell it is anticipated. In sum, only one condition justifies single-framework rule over framework competition, and that is continuous discussion of the theoretical and aim-improvement problems that the framework implicitly or explicitly raises. Gutting's framework thus is no substitute for Maxwell's blueprint.

SUMMARY

In their different ways Lakatos and Feyerabend have sought to unite Popper and Kuhn. Lakatos agrees with Kuhn that scientists seek to extend their theories rather than falsify them, but he rejects the notion that a science or field is, or should be, dominated by a single theory at a time. Science, he maintains, grows through competition among research programs. A research program consists of a hard core of assumptions plus a protective belt of auxiliary hypotheses generated by a heuristic. According to Lakatos, it is rational for a scientist to continue working on a research program so long as it is making and confirming novel predictions.

Feyerabend agrees with Popper that science should expose at least some theories to as much criticism as possible. It should do so, he says, by proceeding counterinductively, by inventing theories to predict facts that will refute established theories. The new theories, however, should be protected from criticism until they have shown how much they can predict. Science should proliferate as many such theories as possible. But Feyerabend denies that science is, or should be, rational. "Anything goes," he says. There are no permanent standards for research to follow, and a scientist is entitled to use propaganda and ad hoc hypotheses to keep a new theory in view until evidence accumulates in its favor.

Where Lakatos reacted to Popper and Kuhn, Maxwell has reacted to Lakatos. Maxwell's theory is largely normative. For him, empirical success alone does not suffice for progress in science. A new theory also must render nature more intelligible. Scientific growth should not depend on competition among research traditions whose hard cores are closed to criticism; it should be guided rationally through the discussion, choice, and improvement of metaphysical blueprints proposing ideas about order in nature. Each blueprint should provide methodological rules and a language for the formulation of testable theories.

Replying to Feyerabend, Gutting has proposed that theory-proliferation can be kept within rational limits if it is guided by a broad conceptual framework (such as the Newtonian) developing through extension and synthesis. Gutting also shows how in practice a Maxwellian blueprint might guide a research tradition.

Of the theories we have considered in the last two chapters, which, if any, should we accept?

In my view, Popper's falsificationism is inadequate as an account of how science actually grows. It also imposes too narrow and stringent a form of rationality in its demand that a theory should be rejected at the first agreed-on refutation.

Feyerabend advances some important and stimulating ideas, but his work is more critical than constructive. His concern is with the overthrow of established theories rather than with showing how theories can guide research over a period of time. Moreover, as a proclaimed anarchist he denies that science can, or should be, purely rational, a view I reject.

Lakatos has proposed a detailed, testable theory of science at the research-tradition level. His analysis of the elements of a research program is stimulating and incisive. But this theory is flawed by his failure to provide either consistent standards for appraising the success of research programs or any method of inventing such programs rationally.

The problems posed by Lakatos are solved in Maxwell's normative theory of aim-oriented empiricism. They are solved, in my view, by Maxwell's move to the higher level of the publicly discussed, aim-including, metaphysical blueprint.

Of the theories of scientific change and rationality that I know, Maxwell's is my first choice. It is broad in scope, closely and power-fully argued, and is in keeping with the purpose of this book, which is to see science in its totality. No other theory provides, as Maxwell's does in principle, for the rational direction of the overall growth of science. But Maxwell still has to show in detail that competition among ideas for blueprints stimulates scientific growth as keenly and critically as rivalry among research traditions or theories. He also has to demonstrate how a blueprint would in practice steer an empirically successful research tradition. One way for Maxwell to do the latter would be to draw on Gutting.

With this chapter I conclude my largely philosophic analysis of the historical movement of science. I have sought to give the reader an insight into scientific growth and the inherent, though inevitably limited, rationality of this growth. But let me emphasize again that in the last two chapters I have discussed the growth of science only so far as it occurs through decisions made by scientists for scientific reasons. Science also is influenced by social and cultural factors outside itself. To understand scientific growth at all adequately, we must take this influence into account, as I shall try to do later.

NOTES

1. Lakatos's views are set forth in "Falsification and the Methodology of Scientific Research Programmes" in *Criticism and the Growth of Knowledge*, ed. Imre Lakatos and Alan Musgrave, pp. 91–196; "History of Science and Its Rational Reconstruc-

tions," in *PSA 1970: In Memory of Rudolf Carnap*, Boston Studies in the Philosophy of Science, vol. 8, ed. Roger C. Buck and Robert S. Cohen (Dordrecht, Holland: Reidel, 1971), pp. 91–136; "Popper on Demarcation and Induction," in *The Philosophy of Karl Popper*, ed. Paul Arthur Schilpp, The Library of Living Philosophers, vol. 14, book 1 (LaSalle, Ill.: Open Court, 1974), pp. 241–73; and "Why Did Copernicus' Research Program Supersede Ptolemy's?" in *The Copernican Achievement*, ed. Robert S. Westman (Berkeley and Los Angeles: University of California Press, 1975), pp. 354--83. His theory of research programs is applied in a number of case studies in *Method and Appraisal in the Physical Sciences*, ed. Colin Howson, which also contains a fine critique by Feyerabend.

2. Lakatos, "Popper on Demarcation and Induction," p. 248.

3. Lakatos, "Falsification and the Methodology of Scientific Research Programmes," pp. 133–34, 144–46, and "Why Did Copernicus' Research Program Supersede Ptolemy's?" pp. 368–69.

4. A classic example is the Lorentz-Fitzgerald contraction hypothesis, which was proposed to explain the failure of the Michelson-Morley experiment to detect an increase in the speed of light traveling in the same direction as the earth's motion and the ether. The hypothesis stated that objects contract at high speeds, and that the interferometer arm had shrunk in proportion to the change in velocity of light, which it had therefore failed to measure. Thus the fact that light apparently traveled at the same speed both with and against the ether did not disprove the existence of the ether.

5. The concept "ad hoc" is more complex than this formulation would suggest, and recently it has been much debated. See, e.g., Lakatos, "Falsification and the Methodology of Scientific Research Programmes," p. 175 (including nn. 2, 3), and "History of Science and Its Rational Reconstructions," p. 103 n. 36; Elie Zahar, "Why Did Einstein's Program Supersede Lorentz's?" in *Method and Appraisal in the Physical Sciences*, ed. Howson, pp. 216–19; Jarrett Leplin, "The Concept of an Ad Hoc Hypothesis," *Studies in History and Philosophy of Science* 5 (1975): 309–45; and Adolf Grünbaum, "Ad Hoc Auxiliary Hypotheses and Falsificationism," *British Journal for the Philosophy of Science* 27 (December 1976): 329–62.

6. Under a spectroscope the heated vapor of every chemical element emits its own characteristic colored line or set of lines depending on the motion of the element's electrons. Each pure colored line has a specific wavelength.

7. Lakatos, "Falsification and the Methodology of Scientific Research Programmes," pp. 146–54.

8. Zahar, "Why Did Einstein's Program Supersede Lorentz's?" p. 262.

9. Lakatos, "Falsification and the Methodology of Scientific Research Programmes," pp. 158–59.

10. *Ibid.*, pp. 138–40.

11. *Ibid.*, p. 159. John Worrall maintains, on the other hand, that Young's versions of the wave program were degenerative and ad hoc ("Thomas Young and the 'Refutation' of Newtonian Optics: A Case-Study in the Interaction of Philosophy of Science and History of Science," in *Method and Appraisal in the Physical Sciences*, ed. Howson, pp. 107–79). Double-slit experiment: In 1802 the British physicist Thomas Young identified the phenomenon of light interference, thus demonstrating the wave nature of light. He observed that when light is passed through two closely set pinholes onto a screen, it splits into two beams, which then recombine to show a pattern of light and dark bands. He concluded that the bands recur because the peaks and troughs of the recombining beams are not always in phase. When two peaks coincide, they reinforce each other, producing a line of light; when a peak and a trough coincide, they cancel each other, yielding a dark line.

12. Lakatos, "History of Science and Its Rational Reconstructions," p. 105.

13. Zahar, "Why Did Einstein's Program Supersede Lorentz's?" pp. 250–62, 271–72.

14. Lakatos, "Falsification and the Methodology of Scientific Research Programmes," p. 153. One of the greatest creative shifts in the history of science was Einstein's introduction of the "principle of equivalence" (i.e., of the equality of inertial and gravitational masses), through which he created the general theory of relativity (Zahar, "Why Did Einstein's Program Supersede Lorentz's?" 263–69).

15. *Against Method*, p. 32.

16. "Problems of Empiricism, Part II," in *The Nature and Function of Scientific Theories: Essays in Contemporary Science and Philosophy*, ed. Robert G. Colodny, University of Pittsburgh Series in the Philosophy of Science, vol. 4 (Pittsburgh: University of Pittsburgh Press, 1970), pp. 292–93.

17. How data are "theory-laden" will be considered in the next chapter.

18. "Against Method: Outline of an Anarchistic Theory of Knowledge," in *Analyses of Theories and Methods of Physics and Psychology*, ed. Michael Radner and Stephen Winokur, Minnesota Studies in the Philosophy of Science, vol. 4 (Minneapolis: University of Minnesota Press, 1970), pp. 22, 26.

19. "Problems of Empiricism, Part II," p. 294.

20. See *ibid.*, p. 307: "The methods of reminiscence to which Galileo appeals so frequently are designed to create the impression that nothing has changed and that we continue describing our observations by means of the old and familiar terms." And p. 313: "For Galileo uses propaganda; he uses psychological tricks. . . . These tricks are very successful. . . . They obscure the fact that the experience [i.e., of regarding all observed motion as relative motion] on which Galileo wants to base the Copernican view is nothing but the result of his own fertile imagination, that it has

been invented. They obscure this fact by insinuating that the new results which emerge are known and concealed by all and need only be called to our attention to appear as the most obvious expressions of the truth."

21. "Problems of Empiricism," in *Beyond the Edge of Certainty: Essays in Contemporary Science and Philosophy*, ed. Robert G. Colodny, University of Pittsburgh Series in the Philosophy of Science, vol. 2 (Englewood Cliffs, N.J.: Prentice-Hall, 1965), pp. 174–75.

22. This phenomenon was first noticed in 1827 by the Scottish botanist Robert Brown, who was observing pollen grains in water. (See ch. 1.)

23. "These universal theories are theories which enable one to say something about anything there is in the world." Feyerabend in "Discussion at the Conference on Correspondence Rules," in *Analyses of Theories and Methods in Physics and Psychology*, ed. Radner and Winokur, p. 246. Examples are Aristotelian physics, Newtonian mechanics, classical thermodynamics, and the special and general theories of relativity. (See *ibid.*, and "Problems of Empiricism, Part II," pp. 294–96).

24. "Problems of Empiricism, Part II," pp. 318–19.

25. Feyerabend, *Against Method*, p. 23.

26. *Ibid.*, pp. 46, 30, 189, 32.

27. For a case study of a revolutionary scientist who did *not* proceed counterinductively but used known methods to solve a known problem and predict known facts, see Gary Gutting on Max Planck's solution of the black-body problem in "Conceptual Structures and Scientific Change," *Studies in History and Philosophy of Science* 4 (November 1973): 212–16.

28. See ch. 2.

29. Nicholas Maxwell, "The Rationality of Scientific Discovery, Part I: The Traditional Rationality Problem"; and "The Rationality of Scientific Discovery, Part II: An Aim-Oriented Theory of Scientific Discovery."

30. Maxwell, "The Rationality of Scientific Discovery, Part I," p. 145.

31. John Archibald Wheeler, *Geometrodynamics* (New York: Academic Press, 1962). In 1972, however, he abandoned this view. See C. W. Misner, K. S. Thorne, and J. A. Wheeler, *Gravitation* (San Francisco: Freeman, 1973), pp. 1203–8.

32. See "Twistors and Particles: An Outline," in *Quantum Theory and the Structure of Time and Space*, papers presented at a conference held in Feldafing, July 1974, ed. L. Castell, M. Drieschner, and C. F. von Weizsäcker (Munich: Carl Hanser Verlag, 1975), pp. 129–45.

33. Maxwell, "The Rationality of Scientific Discovery Part II," pp. 255–64.

34. Conservation laws in physics state that certain properties of an isolated physical system (such as electrical charge, mass-energy, and linear and angular momentum)

do not change in time. Conservation laws exist because, owing to the geometrical symmetry of space and time, laws of nature are symmetrical, that is, invariant or changeless under various symmetry operations such as rotations, translations, and reflections of the spatial and temporal coordinates. Both relativity theory and quantum mechanics involve symmetry fundamentally. The special theory of relativity can be stated as the invariance, or unchanging form, of the laws of physics under a continuous group of symmetry transformations known as the Poincaré group. In quantum mechanics every symmetry produces a conservation law. That is, for every symmetry operation that leaves the laws invariant there is a physical property, such as charge, that remains unchanged during the development of an isolated system.

35. *Ibid.*, p. 284: This case "illustrates with dazzling clarity the thesis that we can only expect the fundamental operations of nature to have a 'simple' form when written in terms of terminology appropriate to the proper blueprint. If formulated in terminology appropriate to some incorrect blueprint, the fundamental laws of nature can be expected to have only an impossibly intricate, complex form."

36. *Ibid.*, pp. 291–93. Maxwell has suggested what a microrealistic quantum theory might be like in his "Toward a Micro-Realistic Version of Quantum Mechanics," *Foundations of Physics* 6 (June 1976): 275–92, and *ibid.* (December 1976): 661–76.

37. Maxwell, "The Rationality of Scientific Discovery Part II," pp. 286–87.

38. Gary Gutting, "Conceptual Structures and Scientific Change."

39. *Ibid.*, pp. 227–28.

40. A book, *The Aims of Science*, is said to be forthcoming.

41. Maxwell, "The Rationality of Scientific Discovery, Part I," p. 150.

42. Gutting, "Conceptual Structures and Scientific Change," pp. 216–21.

43. *Ibid.*, p. 219.

44. *Ibid.*, p. 220.

Chapter 5

A Method of Inquiry

SCIENTIFIC METHOD: THE IDEAL AND THE REAL

To the lay visitor a scientific laboratory hums with efficiency. Not so to most scientists. They know how much of research is trial and error—how much depends on factors other than scientific laws and method. As biologist J. Z. Young says:

> in his laboratory he [the scientist] does not spend much of his time thinking about scientific laws at all. He is busy with other things, trying to get some piece of apparatus to work, finding a way of measuring something more exactly, or making a dissection that will show the parts of an animal or plant more clearly. You may feel that he hardly knows himself what law he is trying to prove. He is continually observing, but his work is a feeling out into the dark, as it were. When pressed to say what he is doing he may present a picture of uncertainty or doubt, even of actual confusion.[1]

Is there a method in this confusion?

There *may* be. Some writers have claimed that all research projects involve the same core activities. But this claim is surely false. Hypothesizing is the essence of theory-construction, yet in ordinary fact-finding no hypotheses are invented. (A hypothesis is a research-guiding conjecture.) Hence, there is no single scientific method, in the sense of a single sequence of research acts exemplified in all varieties of research. Nevertheless, all original research projects, all investigations in which a hypothesis is formed, do involve a common cycle of activities. This cycle is strikingly like the structure of thoughtful problem-solving in everyday life. Consider an example.

In a certain city a new road had been built and the accident

96

rate soared. There was a public outcry and an investigation followed. The investigators began with the most obvious hypothesis—that a new road increases traffic, which increases accidents. But they found that accidents had mounted disproportionately. They then conjectured that on a new road drivers were less careful. But the statistics on other new roads disproved this. So they surmised that the cause was speeding. According to police records, however, fewer drivers had been cited than usual. Had the police been less active? No, the same number had been on duty. Then the investigators noticed that most of the accidents had occurred in only three places on the road. So they recommended new rules for traffic at these points. After this the number of accidents fell below the norm. The problem had been solved.

The process I have described looks tidy, but that is because it is the structure and not the experience of an investigation. This structure often is concealed from the investigator by the acts in which it is embodied—acts that may be inconsistent or thwarted. Take another example.

Both in literature and in life, detectives eat, drink, and sleep hypotheses. They examine the scene of a crime, interview witnesses and suspects, postulate motives, propose scenarios, and test everything against their data. Some detectives can finger a culprit before the evidence is in. It comes to them "naturally," some people say, but perhaps more from experience with similar cases.

A detective's job can be quite as difficult as a scientist's. After working a string of hypotheses, the detective may get his man. It may take him a few days, a few months, a few years. Or he may fail. Most crimes are never solved. Now listen to a scientist: "Nearly all scientific research," says biologist Peter Medawar, "leads nowhere—or, if it does lead somewhere, then not in the direction it started off with. . . . I reckon that for all the use it has been to science about four-fifths of my time has been wasted, and I believe this to be the common lot of people who are not merely playing follow-my-leader in research."[2]

Now of course the detective analogy, like all analogies, breaks down in certain respects. Detectives and scientists have different goals. The detective aims to catch a criminal; the scientist, to contribute to knowledge. Also the techniques of their investigation differ, owing to the kinds of evidence each seeks to obtain. Nevertheless, in both examples we find the same sequence of activities as in scientific research—problem, hypothesis, inference, test, feedback, change of hypothesis, and the sequence repeated. Thus the scientific method is not unique. Scientific inquiry uses more sophisticated knowledge and more refined techniques than those of everyday thoughtful problem-solving, but its rational structure is the same. Let me outline this structure more formally.

While carrying out observations or experiments, or reflecting on current knowledge, the scientist notices something unusual, such as a fact at odds with an established theory or an inconsistency among theories. (Darwin, for instance, noticed 13 species of finches on the Galápagos Islands; Einstein saw that Newtonian mechanics and Maxwell's electrodynamics were inconsistent.) He formulates the disagreement as a problem to be investigated. After more observation or reflection he proposes a solution—a hypothesis that something is the case. He then deduces the implications of this hypothesis, predicting what states of affairs should obtain if the hypothesis is correct. If these states of affairs are observable (i.e., if there are instruments that can detect them), he carries out observations or experiments to collect data on them. He compares the data with his predictions, and if the two sets of statements agree, he considers the hypothesis to that extent confirmed. If they disagree, he has three options: to make further predictions and conduct further tests; to propose another hypothesis, deduce its implications, and test them (a process he may repeat several times); or to abandon the project altogether. If his predictions are confirmed (or if he expects them to be), he writes a draft of his solution, stating his hypothesis, data, and predictions. This is the first research cycle, the cycle of *discovery*. It accommodates all the unforeseen events mentioned above, and it forms the structure of all research projects in which a hypothesis is invented.

It is followed by another, the cycle of *validation*. The scientist now submits his solution to the judgment of his peers. He therefore must relate the solution to established knowledge and show that his arguments and techniques meet the standards of the field. As a rule he presents a preliminary report at a meeting of his fellow specialists and defends it against criticism. Next he writes a formal paper which he sends to a professional journal. His solution then is checked by other scientists for soundness of reasoning, accuracy of calculation, adequacy of evidence to conclusion, and significance of the problem itself. If the solution survives repeated tests, it is accepted as reliable and used in the investigation of further problems.

TYPES OF RESEARCH

The method I have described is used in some scientific research, but not in all. The main types of research are fact-finding, consolidation, extension, reformulation, and theory-creation.

At least half of scientific research consists of fact-finding or gathering data about phenomena already partly known, such as the positions of stars, the specific gravities of materials, wave lengths, electrical conductivities, the boiling points of solutions, and so on.[3] It includes the testing of laws, theories, and hypotheses, and experimenting with new instruments and techniques to see what they will achieve. In such research, hypotheses ordinarily are not invented, so the method I have described is not normally used.

Consolidation consists in developing the implications of a law or theory for the areas in which it is expected to apply. During the eighteenth and early nineteenth centuries, for example, many scientists sought to predict the motions of the moon and planets from Newton's laws of motion and his law of gravitation. In 1846 Antoine Leverrier correctly predicted the existence of the planet Neptune.

Extension is the application of a law or theory to new areas. In the eighteenth century, scientists applied Newton's laws of motion to hydrodynamics and vibrating strings, and in 1905 Einstein used Planck's quantum theory to propose that light travels in photons. For half a century, as we have seen, dedicated physicists have sought to consolidate and extend general relativity.

Reformulation is the revision of a theory to make it clearer, simpler, or more easily applicable. During the eighteenth and nineteenth centuries, a succession of brilliant mathematicians (Euler, Lagrange, Gauss, Hamilton) reformulated Newton's laws and worked

99

out techniques to apply them more widely and precisely. In this century scientists have worked on the mathematical and philosophic foundations of quantum mechanics.

Theory-construction, including the creation of new laws and taxonomies, is the most vital and original form of scientific research. All forms, however, other than fact-finding, entail the invention of hypotheses and hence the use of the method I have described.

Techniques. As a rule this method is used together with special techniques, acquired largely through practice during the scientist's apprenticeship.[4] These techniques may be conceptual (such as algorithms—step by step procedures—for deducing consequences and checking solutions) or empirical (such as procedures for making observations and performing experiments). Each science has its own techniques. Biologists, for example, but not astronomers, use control groups. A scientist can divide a group of rabbits with similar characteristics, treat both sets alike on every characteristic but one, and observe the results. But he cannot do so with stars or galaxies. In chemistry some widely used techniques are solution, filtration, evaporation, distillation, and crystallization. Different branches and specialties also have characteristic techniques. Most organic chemists use spectrometers, whereas physical chemists have only the computer in common. Again, within physical chemistry many specialties are distinguished by the use of a particular instrument: flash photolysis, laser X-ray, photoelectron spectrometry, low-energy electron diffraction, and so on.[5]

All sciences, however, use concrete models. For centuries astronomers have used the orrery, a model of the bodies and movements of the solar system. Today biochemists and molecular biologists employ models of the atomic structure of molecules. Watson and Crick made several such models on the way to their theory of the helical structure of the DNA molecule.

FACTS AND DATA

The scientist observes facts and records them in data. *Facts* are things that happen or subsist; they are events or states. *Data* are symbolic representations of events and states—generally statements recording them. Surprising as it may seem, there is no fact that is not colored by our preconceptions. This can be shown from everyday experience, for what we mostly perceive are objects and processes of definite kinds, not raw sense impressions in "blooming, buzzing, confusion." We interpret sense impressions by means of concepts, and so have perceptions rather than meaningless sensations. Thus perception is essentially interpretive or judgmental. As a contemporary philosopher has said:

> Perception must . . . be understood as the activity of referring a present sense-content to the systematically structured background knowledge of the world; and the successful outcome of this activity is the achievement of recognition. . . . But the fact has not been *given* to us gratis. We have achieved it through a complex activity of schematizing, organizing, reference, and interpretation of the scrutinized contents of primitive sentience.[6]

In science the conceptual schemes that enter into our observations are more theoretical, more exact, and more consciously criticized than those of ordinary life. Hence both facts and data are "theory-laden." In Russell Hanson's words, "The observations and the experiments are infused with the concepts; they are loaded with the theories."[7] This happens because the theories themselves define the concepts in terms of which the data are expressed and the facts are interpreted (see chapter 2). As things are conceived, so are they seen. The theories not only direct our attention to things we had not noticed before, they also influence what we see when we see it. In some cases this is obvious. To a layman a photograph of a bubble chamber is a pretty picture. Only a particle physicist can read the lines and spirals as the tracks and collisions of specific particles. But it also is true of the direct observation of gross physical objects like the sun. As Hanson shows, Tycho Brahe, holding the geocentric theory, and Johann Kepler, holding the heliocentric, saw the same sun differently. Brahe saw the sun rise over the earth, but Kepler saw the earth's horizon fall away from the sun.[8] Or take Kuhn's example of a stone swinging from a fixed support.[9] For an Aristotelian what counted was the coming to rest of the stone in its lowest position. For

101

Galileo what mattered was that the stone repeated its motion and instead of remaining at its lowest point swung to the opposite extreme. Thus, where the Aristotelian saw a stone prevented from falling, Galileo saw a pendulum.

Nevertheless, although the facts are theory-laden, they need not all be loaded by the theory or theories under test. At sunrise Brahe and Kepler saw the sun's orb and the horizon move apart. This fact, laden with a much older theory, was common to them both. Similarly, both Galileo and Aristotle saw a stone swinging. Since rival scientists see at least some facts alike, competing theories can be compared. Moreover, theory-loading does not make theories self-confirming. Theories determine what the facts will be like but not what they will be—what could confirm them, but not what does.

OBSERVATION

The scientist observes much more carefully than the ordinary man. The good scientist looks for the unexpected. Of Charles Darwin, his son wrote: "There was one quality of mind which seemed to be of special and extreme advantage in leading him to make discoveries. It was the power of never letting exceptions pass unnoticed."[10] The same might have seen said of Pasteur. One day, while watching bacteria in a tiny drop of fluid undergoing butyric acid fermentation, he was astonished to see that as the organisms approached the edge of the drop they stopped moving. He guessed that this was because near the air the fluid contained oxygen. What did this fact signify? That these bacteria lived where there was no oxygen. From this insight he leapt to the conclusion that life can exist without oxygen, a condition generally thought to be impossible. This important hypothesis sprang from the observation of an anomaly that few would have noticed.[11]

Pasteur was using a microscope. Instruments enormously increase the range and accuracy of observation. With some instruments, such as the microscope and telescope, we observe phenomena directly. With others what we observe directly is treated as evidence of something unobserved. Take the Fermi Laboratory accelerator near Chicago. This huge circular machine, 4 miles in circumference, accelerates particles to velocities near the speed of light and then slams them together. As the particles collide, they disintegrate into their constituent parts, or at any rate into some of them. The collisions may be photographed in a bubble chamber, a sphere filled with liquid hydrogen. When a beam of particles hurtles into the chamber, a piston moves, releasing the pressure. As though a bottle of champagne had been opened, bubbles form in the liquid hydrogen along the ionized paths of the particles. The tracks revealed by the bubbles are photographed, telling scientists of a world that cannot be observed directly.

Scientific observation is systematic, detailed, and varied. It is made *systematic* by being controlled by a hypothesis or by a precise idea of the phenomenon to be located. It is made *detailed* by using powerful instruments and by concentrating on particular properties of a phenomenon. It is made *varied* by observing the phenomenon under different conditions, or, in an experiment, by varying and holding constant different variables in order to note the results.

The data obtained by observation are expected to be objective, reliable, and precise. Data are *objective*, or intersubjectively testable, in the sense that any suitably trained scientist, performing the same operations, is expected to observe the same facts as those recorded and thus obtain the same data. To this end, data are expressed in the language of physical things (rocks, plants, colors, sounds, weights, pointer readings) rather than in terms of sensations unique to the individual. The data are *reliable* when the facts are given a description that different scientists, observing the facts, can accept. The data also are expected to be *precise*; they should describe a fact so as to differentiate it as much as possible from similar facts. The most objective, reliable, and precise data are quantitative. How do we obtain them?

MEASUREMENT

Instead of describing a set of objects as "many," we may give it a number; instead of calling something long or short, we may note that it is so many meters long. But to do so we need ways of assigning numbers to things. One way is to count. In this case we put the things into a one-to-one correspondence with the natural numbers. When we count from 1 to 20, for instance, we establish that there are as many objects as there are natural numbers from 1 to 20.

But counting only tells us how many items there are in a collection. Measurement, on the other hand, tells us how much of a property is present in an object or event. A property cannot be counted, because it does not consist of discrete units. Instead, numbers and units have to be assigned to it. Measurement is the assignment of numbers to properties according to a rule. The property measured is called a "quantity," and the number representing the amount of this quantity is called its "measure." Thus temperature is a quantity, and 83 degrees is a measure of it.

Measurements of size and weight are more precise than qualitative assessments such as "huge," "tiny," and the like; and specifying colors by wavelengths is more precise than calling them "pink" or "blue." Measurement also enables us to state laws exactly, to assert not merely that one thing depends on another but that changes in one variable property correspond to a precise degree to changes in another.

Units of measurement are numbered to form a *scale*, which (in one sense of this term) is an arrangement of numbers in their natural order along a line. An instrument that carries these numbered units is a measuring scale, such as a tape measure, a thermometer, a spectroscope, or a Geiger counter. The precision of the scale depends on the amount of its smallest unit; the smaller the unit, the more precise the scale.

Since the properties of objects do not naturally possess the characteristics of numbers, some rule must be devised for assigning numbers to them. In fact, several rules have been invented, each providing for the assignment of numbers with different characteristics. These rules also are called "scales."

How are the numbers assigned? The basic measuring operation is to match an object possessing a specific property with a device representing so many standard units of that property. To ascertain the length of a table, we match it with a measuring tape. Sometimes the matching is indirect, as when the temperature of an object is compared with the height of a column of mercury in a thermometer. The height of the column is not the temperature of the column but is related to that temperature by a law stating that the warmer an object is, the higher the mercury will rise in a thermometer placed next to it. The first kind of measurement is called fundamental, the second, derived.[12]

THE RESEARCH PROBLEM

The research cycle begins, however, not with observation or measurement, but with the search for, or location of, a problem.[13] This problem may be empirical, such as the existence of an anomaly to a well-confirmed law or theory. In 1933, for example, Carl D. Anderson in Pasadena found evidence of what looked like a positive electron. Hitherto scientists had recognized only negatively charged electrons and positively charged protons. Yet tracks in the cloud chamber now suggested that a particle existed with the mass of an electron and a double (positive and negative) curvature. This was doubly anomalous. Positive curvature normally implied a mass 1,000 times greater than that of the electron. Moreover, a particle of such mass, following a track with the curvature shown in the photograph, would have a range of 5 millimeters, whereas the track itself was 5 centimeters long. So Anderson posed this problem: Is a positive electron possible? Is this the strange particle predicted by Paul Dirac? He answered, rightly, that it was.

Or the problem may be conceptual. The Copernican theory made a number of assumptions about the motion of bodies that clashed with the established Aristotelian dynamics. One of the strongest arguments against the theory was that it was unsupported by any theory of motion that would justify Copernicus's assumptions about the motion of the earth (e.g., that the earth rotated on its axis once every 24 hours). Recognizing this conflict between the two schemes, Galileo created a new dynamics of relative motion that was compatible with the Copernican theory.

Often, however, the scientist begins research by looking for a problem. He may choose an area that is well developed theoretically and hence full of leads for further investigation; or he may go where there has been a sudden rise in the rate of empirical discovery; or he may simply have a hunch that an area is rich in possibilities. This seems to have been why physicist Bruno Rossi of MIT joined the new field of X-ray astronomy, where he discovered (1962) the first X-ray source outside the solar system, Scorpio X-1. He describes his motivation as follows:

> The initial motivation of the experiment which led to this discovery was a subconscious feeling for the inexhaustible wealth of nature, a wealth that goes far beyond the imagination of man. . . . More likely it was inborn and was the reason why, as a young man, I went into the field of cosmic rays. In any case, whenever technical progress [in this case, space science techniques] opened a new window into the surrounding world, I felt the urge to look through this window, hoping to see something unexpected.[14]

THE HYPOTHESIS

Having formulated his problem, the scientist looks for a hypothesis. A hypothesis is a conjecture that something is the case. It normally is expressed in a statement or set of statements from which conclusions can be drawn about what else would be the case if this is. It often takes the form, "If A is true, then B might (should, will) follow."

What are characteristics of a good hypothesis? First, it should account for the known facts. (Nevertheless this qualification sometimes may be disregarded, since a scientist proposing a revolutionary new theory may have to ignore some of the accepted facts while looking for new facts of his own.) Second, it must be precise enough to yield testable predictions. As such it is valuable even when incorrect, for if it can be decisively refuted, it can be eliminated as a possible solution. Third, it should predict some fact or facts hitherto unknown. Einstein, for instance, deduced three predictions from his theory of general relativity: the deflection of light in the sun's gravitational field, the motion of Mercury's perihelion, and the red shift of light from distant stars. The first two predictions were confirmed in due course, and the third agreed with existing data. But many scien-

106

tists did not consider the third prediction novel, since the disagreement with Newton had been known for almost a century. However, it recently has been argued that a new fact is better regarded as any fact that a theory has not been invented to explain.[15] By this criterion, Mercury's perihelion and the Michelson-Morley experiment were novel facts for the general and special theories of relativity respectively.

FROM PROBLEM TO HYPOTHESIS

Sometimes a scientist leaps to a hypothesis almost as soon as he sights a problem. In 1895 Wilhelm Röntgen noticed a greenish glow coming from a cathode-ray tube in his laboratory. Thinking that the glow might be caused by ultraviolet rays, he put a fluorescent screen near the tube. It lit up. He then put the tube in a cardboard box. Still the screen lit up. This showed that the radiation could not be ultraviolet light, which does not penetrate cardboard. Röntgen reflected: the rays passed through the glass tube, the cardboard box, and the air to illuminate the fluorescent screen; they therefore must be an unknown form of invisible light, and if so, they must cast a shadow. On an impulse, Röntgen put his hand in front of the screen. To his amazement he saw not the shadow but the skeleton of his hand, the flesh and skin forming a faint, grayish fringe. He realized immediately that he was dealing with an entirely new kind of radiation. After carrying out further experiments, he published a paper describing the properties of these "X-rays," as he called them. And "X-rays" they have remained. Poor Röntgen—to have had the measuring units named in his honor, but not the rays!

A complex, revolutionary hypothesis, on the other hand, may take some time to form. During his five years on H.M.S. *Beagle* (1831–36), Darwin, as ship's naturalist, amassed a store of evidence about the plants, animals, and geological strata of South America. But he was interested mainly in geology. His visit to the Galápagos Islands, where he saw small variations in the species of birds and tortoises from island to island, shook his belief that species were immutable. In 1835 he proposed a theory of coral reefs that resembled his later theory of evolution. But he did not become an evolutionist until nearly two years after his return to England.[16] At this point he put forward two different evolutionary hypotheses, only to reject them. Then he began to search for the specific cause or causes of evolution. The idea that natural selection is the cause struck him over a year later while he was reading Malthus. Darwin described this insight in his *Autobiography*:

> In October 1838, that is fifteen months after I had begun my systematic enquiry, I happened to read for amusement Malthus on *Population*, and being well prepared to appreciate the struggle for existence that everywhere goes on from long-continued observation of the habits of animals and plants, it at once struck me that under these circumstances favorable variations would tend to be preserved and unfavorable ones to be destroyed. The result of this would be the formation of new species. Here, then, I had at last got a theory by which to work.[17]

Darwin did not write a draft of his hypothesis until four years later, when he produced an abstract of 35 pages. Two years afterward he enlarged this abstract to 230 pages. For the next 14 years he discussed the theory with correspondents but did not announce it publicly. Then he received a paper from Alfred Russel Wallace giving a brief statement of the theory, which Wallace had thought up independently. A joint paper by Darwin and Wallace was read before the Linnaean Society of London on July 1, 1858, and then published. Darwin's book, *The Origin of Species*, appeared on November 24, 1859, and immediately booksellers bought up the entire edition of 1,500 copies.

Why did Darwin wait so long to announce his theory? Partly because he wished to perfect the theory, but mainly because he feared persecution for advocating a view that might be stigmatized as a rejection of the Biblical account of Creation. As Darwin knew, Galileo, at 70, had been forced by the Inquisition to renounce the Copernican theory, and Giordano Bruno, an earlier supporter of that theory, had been burned at the stake. Although he did not expect so grim a fate, Darwin nonetheless hesitated to defy public opinion.

After proposing a firm hypothesis, what does the scientist do with it? As a rule he tests the hypothesis by making observations or carrying out an experiment. The test often is inconclusive. So, with the aid of data provided by the test, the scientist may refine his hypothesis through a series of trials (i.e., reformulations and tests), each trial furnishing data with which he can make the hypothesis more precise. Alternatively, instead of refining a single hypothesis, he may test a series of different ones. Trying to calculate the orbit of the planet Mars, Kepler began with the hypothesis that it was circular. He found this hypothesis refuted by the facts, but tried it again, only to find it refuted more decisively still. So he added a bulge to one side of the circle and made it an ovoid. Since this hypothesis turned out to be self-contradictory, he modified the ovoid so that it began to resemble an ellipse. Finally, he proposed that the form of a planet's orbit is a perfect ellipse.

Another tactic is to propose a number of hypotheses at the outset and eliminate them successively until only one remains. Alternatively, the scientist may fit them one by one into a general scheme. Thus Darwin took a first hypothesis from domestic breeding, a second from the struggle for existence and the natural selection of wild varieties, a third from the ramification of species from common progenitors as shown in the geological record, and a fourth from the geographical distribution of species. He combined these hypotheses to form his theory of the evolution of species.

In the course of a research project the scientist gathers data as well as invents hypotheses. After most tests he uses the data to modify or replace his hypothesis. Let us see how Lord Rayleigh, for example, proceeded through a succession of hypotheses and tests to the discovery of the gas argon.[18] In 1892 Rayleigh discovered that atmospheric nitrogen is one-half percent heavier than nitrogen prepared from chemical compounds. Why was this? His first hypothesis was that when nitrogen is chemically prepared, it becomes mixed with a light gas such as hydrogen. As a test he introduced hydrogen into nitrogen, but found the density unaffected. Out went that hypothesis. The alternative was that there is an unknown heavy gas in the atmosphere. The way to test this was to eliminate the real nitrogen from the atmosphere and see if anything was left. So he mixed nitrogen with oxygen and sent electric sparks through the mixture. (Nitrogen and ox-

109

ygen combine, resulting in a compound that can be removed.) The experiment dragged on for almost two weeks. The sparking apparatus kept stopping, and Rayleigh would doze in an armchair in an adjoining room until late at night with a telephone near him to relay the noise of the instrument. When the noise stopped, he woke and adjusted the apparatus. Eventually a small residue was left. Was this hydrogen or nitrogen? He performed a couple of experiments to disprove the first possibility. First he passed atmospheric nitrogen over red-hot magnesium, leaving a small residue, which turned out to be heavier than nitrogen and thus heavier than hydrogen. Then he carried out an atmolysis of air—a process in which a mixture of gases is leaked through a porous pot—and found a small residue, again heavier than nitrogen. So he proposed that there exists a hitherto unknown gas, argon. He confirmed this hypothesis by carrying out a number of tests eliminating the possibility that the gas was hydrogen. Argon also was found to have a different spectrum than nitrogen and to be two and a half times as soluble in water. This last finding suggested that there should be more argon than nitrogen in rainwater, which tests again confirmed.

REASONING AND HYPOTHESIS-FORMATION

How does the scientist reason when he forms a hypothesis? Many hypotheses seem to arrive in a flash of intuition. Here is a typical case. In 1934 Enrico Fermi made a crucial observation that was to lead to the splitting of the atom. He discovered that a beam of neutrons can destabilize the nuclei in a target far more effectively if the neutrons are first slowed down by passing through a moderator such as paraffin. Fermi later described this moment to astrophysicist Subrahmanyan Chandrasekhar:

> We were working very hard on the neutron induced radioactivity and the results we were obtaining made no sense. One day, as I came to the laboratory, it occurred to me that I should examine the effect of placing a piece of lead before the incident neutrons. And instead of my usual custom, I took great pains to have the piece of lead precisely

110

machined. I was clearly dissatisfied with something: I tried every "excuse" to postpone putting the piece of lead in its place. When finally, with some reluctance, I was going to put it in its place, I said to myself, "No: I do not want this piece of lead here; what I want is a piece of paraffin." It was just like that: with no advanced warning, no conscious, prior, reasoning. I immediately took some odd piece of paraffin I could put my hands on and placed it where the piece of lead was to have been.[19]

Struck perhaps by incidents such as this, Karl Popper and the logical empiricists say that hypothesis-formation is fundamentally nonrational. As Popper puts it, "The question how it happens that a new idea occurs to a man—whether it be a musical theme, a dramatic conflict, or a scientific theory—may be of great interest to empirical psychology, but it is irrelevant to the logical analysis of scientific knowledge."[20] The scientist allegedly does not reason to a hypothesis but only reasons from it. Testing and appraising a hypothesis is a logical procedure and is the scientist's essential task.

But this view almost certainly is mistaken. First, it is an empirical claim with no evidence to support it. Second, even if at some point in the process of invention the scientist must rely on intuition, at other points he may be guided by rational considerations for which rules can be provided. Third, though there may not be a strict "logic" of discovery, there almost certainly are rational principles or "rules of strategy" that scientists follow in forming and pursuing hypotheses. These rules may be codified and formulated as an explicit rationale of discovery. Fourth, and most importantly, intuition, so called, probably is the condensation of one or more lines of rational thought into a single moment of insight. In this moment the mind assembles premises and passes from them to a conclusion, which is the part of this process that it remembers.[21] For example, in 1928 Alexander Fleming's mind probably worked like this: (1) a mold has accidentally fallen into my culture dish; (2) the staphylococcal colonies near it have failed to grow; (3) a secretion from the mold must have killed them; (4) I remember a similar event once before; (5) if I could isolate this secretion from the mold, it could be used to kill staphylococci that cause human infections.

111

We must distinguish, moreover, between movement along a logical or rational structure and the structure itself. No one claims that because a deductive argument is grasped in a moment of insight, the argument has no logical structure. Whether or not one proceeds by conscious steps from premises to conclusion, and whether or not some steps are omitted, the logical structure that relates premises to conclusion is unaffected.[22] So, too, in science, much hypothesis-formation probably has a logical structure that we have overlooked because we have attended to the moment of insight itself rather than to the materials from which the insight came. These materials doubtless were the object of a long process of subconscious reasoning, of which the insight itself was only the culmination. We have to find in the materials themselves a rational structure from which, together with the scientist's own notes and comments, we can reconstruct the process of reasoning, at times subconscious, that led to the formation of the hypothesis. When the scientist invents a hypothesis, he almost certainly reasons subconsciously from premises to a conclusion. If we do not recognize this, it is probably because we know so little of the informal logic of everyday rational thought, of which scientific thought is a refinement. Thus philosopher Peter Caws declares:

> In the creative process, as in the process of demonstration, science has no special logic but shares the structure of human thought in general, and thought proceeds, in creation as in demonstration, according to perfectly logical principles. . . . The logic of scientific discovery, whose rigorous formulation is yet to be achieved . . . will similarly prove to be a refinement and specialization of the logic of everyday invention. The important thing to realize is that invention is, in the strictest sense, as familiar a process as argument, no more and no less mysterious. Once we get this into our heads, scientific creativity will have been won back from the mystery-mongers.[23]

SOME FORMS OF REASONING

I now want to consider several modes of reasoning leading to the formation of hypotheses that have been described by philosophers of science.[24] These are retroduction, hypothetico-deduction, induction, and reasoning by analogy.

In the case of *retroduction* (R-D), the scientist encounters an anomaly and then seeks a hypothesis from which the existence of the anomaly can be deduced. Thus he reasons back from the anomaly to a hypothesis that will explain it. The form of the inference is this: An anomalous fact A is observed; A would be explicable if hypothesis H were true; hence, there is reason to think that H is true. Kepler, for instance, reasoned retroductively to his hypotheses about the orbit of Mars. He began by proposing that Mars moves in a perfect circle. But he found that the predictions he deduced from this hypothesis conflicted with the data of the Danish astronomer Tycho Brahe. The data, therefore, appeared to be inconsistent with the hypothesis of circular motion. So he assumed that the data were correct and sought to explain them by proposing the hypotheses we have mentioned.[25]

Instead of reasoning to a hypothesis from data, the scientist may begin with a hypothesis and deduce conclusions—general statements or particular predictions—from it. This is *hypothetico-deduction* (H-D). Einstein reasoned hypothetico-deductively in constructing his special theory of relativity. He was committed to two fundamental principles: relativity (there is no absolute reference-frame; all motion is relative to an observer) and operational definition (scientific concepts should be defined in terms of observable phenomena). From the first principle he derived the paradoxical conclusion that the speed of light is constant, and found this conclusion confirmed by the Michelson-Morley experiment. From the second he deduced the paradoxical conclusion that measurements of simultaneity and distance are relative, and he found this conclusion incorporated in the Lorentz transformations (fundamental equations proposed by the Dutch physicist Hendrik Lorentz).[26]

The scientist reasons *inductively* when he infers a general regularity from statements of particular instances. Early in the nineteenth century the French scientist Joseph Gay-Lussac reasoned inductively to the law that gases combine in simple ratios. He carried out experiments with various gases such as hydrogen and oxygen, fluoboric acid gas, and ammonia. From the fact that hydrogen and oxygen combine in a simple ratio and that various acid gases do so when combined with ammonia (and from certain theoretical assumptions, including the notion that gases, by virtue of their molecular structure, should obey simple laws), Gay-Lussac concluded that *all* gases combine in simple ratios.[27]

Analogical reasoning is employed when the scientist arrives at a hypothesis by seeing an analogy between apparently unrelated phenomena. Darwin reasoned to part of his idea of natural selection from an analogy between population pressure among human beings (Malthus) and the survival of species in nature. Kekulé reasoned to his theory of the ring structure of the benzene molecule when he perceived an analogy between a snake holding its tail in its mouth and the arrangement of the carbon atoms in the benzene molecule.

TESTING A HYPOTHESIS EXPERIMENTALLY

When the scientist has formulated his hypothesis, he tests it by deducing its implications in the form of predictions and comparing them with the results of observations or experiments.

The reasoning behind the experimental test of a hypothesis is as follows. When the scientist seeks to establish a connection between two sets of events, he usually tries to show that one of these sets is the cause of the other. That is, he seeks to demonstrate that an event of one sort, A, is always accompanied by an event of another sort, B, and that an instance of B never occurs unless an instance of A also occurs. A, then, causes B when A must be present if B is to happen and when, with A absent, B never happens.

To show that one event is the cause of another is by no means easy. Any event, B, in nature usually occurs in combination with so many other events that it is hard to tell which of them is the cause of B and which accompany B by chance. One way of finding out is to create a situation in which we ourselves control the accompanying events (or conditions). We then can manipulate them one after another to ascertain which produces B and which do not. To do this, we produce a change in the condition we think is the cause of B, while keeping all other conditions from changing. If we then observe

114

a change in the event, B, that follows, we may attribute it to the change that we ourselves have produced. This is our first experiment. We may then carry out a second experiment by varying some condition that we think has no significant influence on B, while holding unchanged the condition that we think produces B. If we then observe no significant change in B, we infer that B is affected only by a change in the original condition and not by a change in the other.

When a scientist conducts an experimental test, he first deduces what his hypothesis implies for a given experimental situation and then manipulates the situation to see whether he is right. Consider one of the most famous of all experimental tests, the experiment in vaccination carried out at Pouilly le Fort, France, by Louis Pasteur in 1881. Pasteur wished to test the hypothesis that vaccinating an animal with attenuated (weakened) anthrax bacteria would make the animal immune to the disease of anthrax itself. Given 60 sheep by the local agricultural society, he divided the animals into three groups: (1) a control group of 10 sheep that were to receive no treatment whatever; (2) an experimental group of 25 that would be vaccinated and then inoculated with a highly virulent culture of the anthrax germ; and (3) another group of 25 that would not be vaccinated but would receive the same virulent dose of the germ. The experimental group would be vaccinated twice with anthrax bacteria of decreasing attenuation, at intervals of 10 to 15 days, and would be injected with a virulent dose of the germ 12 to 15 days later. Pasteur predicted that the 25 vaccinated sheep would all survive, and the 25 unvaccinated ones would die. The survivors would then be compared with the 10 control sheep to show that vaccination had done them no harm.

The first vaccinations were carried out on May 5 before a large crowd and were followed later by the second set of vaccinations and the administration of the germ itself. On June 2 Pasteur arrived to see the results. His predictions were fulfilled to the letter,[28] as an eyewitness has described:

> When Pasteur arrived at two o'clock in the afternoon . . . accompanied by his young collaborators, a murmur of applause arose, which soon became loud acclamation, bursting from all lips. Delegates from

the Agricultural Society of Melun, from medical societies, veterinary societies, from the Central Council of Hygiene of Seine et Marne, journalists, small farmers who had been divided in their minds by laudatory or injurious newspaper articles—all were there. The carcasses of twenty-two unvaccinated sheep were lying side by side; two others were breathing their last; the last survivors of the sacrificed lot showed all the characteristic symptoms [of anthrax]. All the vaccinated sheep were in perfect health. . . . The one remaining unvaccinated sheep died that same night.[29]

Experiments also may be carried out for fact-finding purposes with no hypothesis involved. One set of such experiments led, quite unexpectedly, to Rutherford's theory of the atomic nucleus. One day in 1909 Rutherford had a student, Ernest Marsden, try scattering alpha particles through a large angle, because Rutherford did not think it could be done. He described the result in one of his last lectures:

Then I remember two or three days later Geiger [an associate] coming to me in great excitement and saying, "We have been able to get some of the alpha particles coming backwards. . . ." It was quite the most incredible event that has ever happened to me in my life. It was almost as incredible as if you fired a 15-inch shell at a piece of tissue paper and it came back and hit you.[30]

But experimentation is not a *sine qua non* of scientific testing. Often it is physically impossible for the scientist to manipulate the circumstances of the phenomenon he wishes to explain or to do so without distorting them. On other occasions an experiment is not only physically impossible but logically inappropriate. The scientist may wish to explain some past event such as the event or events indicated by the presence of certain fossils in a stratum. Since this event is nonrecurring, it cannot be repeated in an experiment.

Some hypotheses cannot be tested decisively even by observation. Take Darwinian evolution. Although there is much evidence that species evolve, evolution itself is almost impossible to observe, for a variation only establishes itself over the course of many generations, and we cannot be around to watch the entire process. Nevertheless, the theory is taken to be well confirmed, not because it is decisively testable but because it unifies and renders intelligible a great many data that could not be understood without it.

A single successful prediction serves as a first confirmation of a hypothesis but does not make it reliable. That status normally is achieved only after the hypothesis has been tested and confirmed by a variety of scientists under a variety of conditions. A hypothesis, for example, that an agent such as nicotine or cyclamate is cancer-producing usually is tested in a number of laboratories against a range of animal species to determine whether the hypothesis is applied to one species, or several, or all. If the hypothesis is confirmed by a variety of tests, it is regarded as reliable; and scientists may then begin to look for a mechanism to account for the correlation observed.

THE SCIENTIFIC METHOD IN PERSPECTIVE

The method I have described may seem hopelessly ideal. To the scientist, trying one hypothesis after another, the quest for a solution often feels like a series of setbacks leading nowhere. Yet the process of hypothesis, inference, test, and feedback is going on all the same. The scientist's moods may change, and so may his ideas, but not the essence of what he is doing. Looking back from the solution, the way through a problem may seem devious, but it is not devious in relation to the ignorance from which the scientist started. The method we have described is precisely that of feeling one's way in uncertainty.

I am not surprised that the scientist often is unaware of the method he uses, for as I have said, it is only an extension and refinement of the process of investigation followed in everyday life. As Max Planck explains, "Scientific reasoning does not differ from ordinary everyday thinking in kind, but merely in degree of refinement and accuracy, more or less as the performance of the microscope differs from that of the everyday eye."[31] The scientist generally tries harder than the layman to screen out personal prejudice and check for possible error. He seeks to make his assumptions explicit and attends to the work of others in his field. He reports his findings more accu-

117

rately and makes predictions that can in principle be tested precisely. In all these respects he improves on the layman but does not eclipse him. In fields where laymen have experience they may have more insight than highly trained scientists. Farmers or fishermen, for instance, often can predict local weather more accurately than meteorologists (in part, of course, because meteorology has not yet become as exact a science as physics).

SUMMARY

By "scientific method" we mean the rational structure of those scientific investigations in which hypotheses are formed and tested. This structure is much like that of everyday thoughtful problem-solving. Hypothesis, inference, test, and feedback are the core of the structure. The scientist usually begins by noticing an anomalous fact or an inconsistency in theory and posing the discrepancy as a problem. After further exploration he proposes a hypothesis from which he deduces predictions. As a rule he tests the predictions and publishes the hypothesis if he finds them confirmed. If they are refuted, he usually alters the hypothesis, or invents another one, and tries again. This process is self-corrective. By eliminating incorrect hypotheses, the scientist narrows the search for the correct one.

This method is combined with general operations such as observation and measurement, and with various techniques differing from specialty to specialty. Scientific observation, often controlled by a hypothesis and aided by instruments, is more systematic and precise than its everyday counterpart. The data obtained by measurement and observation usually are theory-laden, and they are objective to the extent that they can be replicated by suitably qualified scientists. Scientific research, however, is focused as a rule on problems.

The crucial step in the research cycle is the invention of a hypothesis. The ideal hypothesis is precise and testable, accounts for the known facts, and predicts at least one new fact. The scientist usually tests his hypothesis by deducing its implications and then carrying out observations or experiments to see whether the implications correspond with the facts. Sometimes the scientist will run through a

118

series of hypotheses and tests until he finds a hypothesis that he regards as satisfactory. Many hypotheses are tested experimentally. An experiment enables the scientist to manipulate the conditions accompanying a phenomenon until he discovers which of them cause it.

Not a few hypotheses are born in a moment of insight. Does this mean that the process of hypothesis-formation is nonrational? Not at all. Intuition seems to be only the condensation of a process of reasoning that normally would take longer and that in principle, if not always in practice, can be reconstructed later. Indeed, several modes of reasoning to a hypothesis have been identified: retroduction, hypothetico-deduction, induction, and reasoning by analogy.

The scientific method is not only intrinsically rational; it is a refinement of everyday reasoning. The scientist has received a more specialized training than the layman, but his thinking is not fundamentally different. In the next chapter we shall look at the results of applying this method—the intellectual constructs of science, from data to theories.

NOTES

1. J. Z. Young, *Doubt and Certainty in Science: A Biologist's Reflections on the Brain* (Oxford: Clarendon Press, 1951), p. 2.

2. Peter Brian Medawar, *Induction and Intuition in Scientific Thought*, pp. 31–32.

3. Thomas S. Kuhn, *The Structure of Scientific Revolutions*, p. 25.

4. On the "craft" character of techniques, see Jerome R. Ravetz, *Scientific Knowledge and Its Social Problems*, pp. 101–03, 173–75.

5. Stuart S. Blume and Ruth Sinclair, "Aspects of the Structure of a Scientific Discipline," in *Social Processes of Scientific Development*, ed. Richard Whitley (London: Routledge & Kegan Paul, 1974), p. 228.

6. Errol E. Harris, *Hypothesis and Perception: The Roots of Scientific Method* (New York: Humanities Press, 1970), p. 288.

7. Norwood Russell Hanson, *Patterns of Discovery*, p. 157.

8. *Ibid.*, p. 23: "Tycho sees the sun beginning its journey from horizon to horizon. He sees that from some celestial vantage point the sun (carrying with it the moon and planets) could be watched circling our fixed earth. Watching the sun at dawn through Tychonic spectacles would be to see it in something like this way. Kepler's visual field, however, has a different conceptual organization. . . . But Kepler will see the horizon dipping, or turning away, from our fixed local star."

9. *Structure of Scientific Revolutions*, pp. 118–19.

10. Francis Darwin, *The Life and Letters of Charles Darwin* (London: Murray, 1888). Quoted by W. I. B. Beveridge, *The Art of Scientific Investigation*, p. 103.

11. Beveridge, *Art of Scientific Investigation*, p. 97, and René J. Dubos, *Louis Pasteur: Free Lance of Science* (Boston: Little, Brown, 1950), pp. 134–36.

12. A good account of measurement is found in Norman Campbell, *What Is Science?* ch. 6.

13. Michael Polanyi, "Genius in Science," *Encounter* 34 (January 1972): 118: "All true scientific research starts with hitting on a deep and promising problem, and this is half the discovery."

14. Bruno B. Rossi, in *Bulletin of the American Academy of Arts and Sciences* 30 (February 1977): 17. Sometimes the unexpected occurs, and a problem is solved, without forethought. In Berlin Robert Koch isolated the tuberculosis germ quite accidentally after months of hard, frustrating work. Instead of throwing away some cultures he thought were useless, he absentmindedly put them in a warmer, only to remember to remove them a few days later. To his astonishment, the tuberculosis germ had grown and could in principle be isolated. However, actually isolating it took a long time and much knowledge and skill.

15. Elie Zahar, "Why Did Einstein's Programme Supersede Lorentz's?" in *Method and Appraisal in the Physical Sciences*, ed. Colin Howson, pp. 216–19; and Imre Lakatos and Elie Zahar, "Why Did Copernicus' Research Program Supersede Ptolemy's?" in *The Copernican Achievement*, ed. Robert S. Westman (Berkeley and Los Angeles: University of California Press, 1975), p. 376.

16. *The Autobiography of Charles Darwin*, with original omissions restored, ed. with appendix and notes by Nora Barlow (New York: Norton, 1958), p. 120.

17. Here I follow Howard E. Gruber, *Darwin on Man*; e.g., pp. 100–105.

120

18. There is a fascinating account in the *Life of Lord Rayleigh* by his son Robert John Strutt (London: Edward Arnold, 1924), ch. 11.

19. S. Chandrasekhar, "Remarks on Enrico Fermi," in *The Physicist's Conception of Nature*, Symposium on the Development of the Physicist's Conception of Nature in the 20th Century, ed. Jagdish Mehra (Dordrecht, Holland, and Boston: Reidel, 1973), p. 801.

20. Karl R. Popper, *The Logic of Scientific Discovery*, p. 31; see Carl G. Hempel, *Philosophy of Natural Science* (Englewood Cliffs, N.J.: Prentice-Hall, 1966), p. 15.

21. See Abraham Kaplan, *The Conduct of Inquiry* (San Francisco: Chandler, 1964), p. 14: "What we call intuition is any logic-in-use which is (1) preconscious and (2) outside the inform■■ ■■■■■■■ for which we have readily available reconstructions."

22. According to Peter Caws, "What requires analysis is the structure of the process, not its particular embodiment in a particular individual. The movement from premises to conclusion, by which one grasps a deductive argument, is equally sudden and intuitive. But the intuitive, sudden nature of the mental act of moving from premises to conclusion and grasping their deductive connection does not affect the nature of that connection in its own right, it does not alter the logical structure of the relationship between premises and conclusion" ("The Structure of Discovery," *Science* 166 [December 12, 1969]: 1375).

23. *Ibid.*, p. 1380.

24. For other accounts of the rationality of hypothesis-creation, see Michael Polanyi, *The Tacit Dimension* (Garden City, N.Y.: Doubleday, 1966), pp. 17–25, and "Genius in Science," *Encounter* 38 (1972): 43–46; Bernard Lonergan, *Insight* (London: Longmans, Green, 1957), passim; Herbert A. Simon, *The Sciences of the Artificial* (Cambridge: MIT Press, 1969), ch. 3, and "Does Scientific Discovery Have a Logic?" *Philosophy of Science* 40 (December 1973): 471–80.

25. Norwood Russell Hanson, *Patterns of Discovery* (Cambridge: Cambridge University Press, 1963), ch. 4, and "Retroductive Inference," in *Philosophy of Science: The Delaware Seminar*, ed. Bernard Baumrin (New York: Wiley, 1963) 1:21–37 (a revised version of the first account).

26. Gary Gutting, "Einstein's Discovery of Special Relativity," *Philosophy of Science* 39 (March 1972): 61–62.

27. Peter Achinstein, *Law and Explanation in Science* (London: Oxford University Press, 1971), chs. 6, 7.

28. Although Pasteur is credited with the discovery of immunization against anthrax, his vaccine alone would not have immunized the sheep at the farm, because he would not have been able to keep the temperature at 45 degrees Centigrade long enough for the vaccine to work. Much against his will, Pasteur's assistants

secretly invented a chemical that did the work of heat. Pasteur finally allowed them to use the chemical if they promised not to reveal what they were doing until well after the experiment had been performed. In Berlin, however, Robert Koch *deduced* that a chemical must have been used, for, like Pasteur, he realized the need to keep the temperature high enough to kill the spores without killing the bacteria. Hence he, too, announced what really had been done.

29. Quoted by Magnus Pyke, *The Boundaries of Science* (Harmondsworth, Middlesex: Penguin, 1963), pp. 82–83.

30. Quoted by E. N. da C. Andrade, *Rutherford and the Nature of the Atom* (Garden City, N.Y.: Doubleday Anchor, 1964), p. 111.

31. Max Planck, *Scientific Autobiography and Other Papers*, trans. Frank Gaynor (New York: Greenwood Press, 1968), p. 88. See Caws, "The Structure of Discovery," p. 1379: "The difference between his [the creative scientist's] logic and ours is one of degree, not of kind; we employ precisely the same methods, but more clumsily and on more homely tasks."

Chapter 6

From Data to Theories

THE STRUCTURE OF SCIENTIFIC KNOWLEDGE

Largely by using the method we have examined, the natural sciences have amassed a body of knowledge that is far beyond the grasp of a single mind. Each science studies some aspect of the world. The physical sciences—physics, chemistry, astronomy, the earth sciences—investigate inanimate matter. Physics and chemistry are experimental, laboratory sciences, while astronomy and the earth sciences rely mainly on observation. Chemistry differs from physics in studying the properties of particular kinds of matter rather than the properties of matter in general. Because they focus on a comparatively small number of properties, physics, chemistry, and astronomy have been able to frame mathematical laws and theories and test them precisely.

The life sciences, such as biology, botany, zoology, and physiology, deal with a greater wealth of phenomena. A representative sample of their problems might include the behavior and distribution of animal populations, the taxonomic relations of whales, the functioning of the central nervous system, and the replication of DNA molecules. Owing to their hierarchical structure, biological entities (whether cells or organisms) tend to vary more than their counterparts in physics and chemistry. Because of the sexual process, they differ in their genetic endowment. Hence they are much less predictable. Consider how difficult it it to make predictions from experiments in population genetics. As Theodosius Dobzhansky points out:

> Suppose that the races [of fruitflies that are] crossed differ only in 100 genes. Elementary genetics teaches that with 100 gene differences,

123

3^{100} gene combinations are possible in the offspring. This is an enormous number—many times larger than the total number of Drosophilae which ever existed. Only a negligibly small fraction of the possible gene combinations are realized in any one experimental population, or in all the populations combined. . . . The indeterminacy of our experiments is caused ultimately by the tremendous efficiency of the sexual process in the creation of new genetic endowments.[1]

The biologist, then, rarely can attain the precision of the physicist in studying simpler and more stable things; instead, he describes a much greater range of natural phenomena in greater detail.

There also is a rapidly growing number of intersciences. Astrophysics, for instance, which uses physics to study the phenomena of the heavens, is one of the liveliest areas in science. During the past two decades alone, astrophysicists have discovered quasars, pulsars, and X-ray stars, and have worked out the properties of black holes. Molecular biology is the most dynamic of the life sciences and, through its contribution to the new technology of genetic engineering, the most controversial. Molecular biologists use chemistry, genetics, and physics (e.g., X-ray diffraction analysis) to investigate the structure and interactions of large molecules. Other intersciences include earth and planetary physics, biophysics, biochemistry, and physical chemistry.

Scientific knowledge is expressed in statements, and sets of statements, of four main kinds: observation reports, classification schemes, laws and generalizations, and theories. Much of current physics is organized deductively by the theories of quantum mechanics and relativity, each theory and the statements it entails forming a single logical system. All of chemistry can in principle be derived from quantum mechanics, which has shown that the properties of the 106 kinds of atoms are a consequence of the interaction between the atomic nucleus and its circling electrons. At present, however, owing to mathematical difficulties, only the behavior of the simplest molecules can be deduced directly from quantum mechanics.[2]

Because their subject matter is more complex, neither the earth

124

sciences nor the life sciences have achieved this degree of logical structure. The theory of evolution, for example, with population genetics at its core, loosely organizes the knowledge produced by a variety of less comprehensive sciences, such as systematics (which studies the distribution of living things), morphology (the study of their structure), paleontology (the study of organisms long dead and fossilized), and embryology (the study of the development of living things). Much of this ensemble is only "sketched in," many statements being linked not by logical implication but by informed guesses. Such statements are not strictly deduced from others. In stead, it is thought that they "might follow" from others or that it is reasonable to hold them "in the light of" others. Nevertheless, evolutionary theories probably will be axiomatized more fully in due course.[3]

THE SEARCH FOR ORDER

Science seeks not merely to record particular facts but to discover regularities among them. There is an order in nature, in the sense that many phenomena have common features. When we study processes that occur under a wide range of conditions, we often find that within this range certain features remain constant. We find, for example, that objects released in midair in many ways and in many weathers invariably fall to the ground. Investigating further, we discover that if air resistance is ignored, the acceleration of these objects is constant. Similarly, we observe that in many climes and countries plants grow from seed to harvest, and that the fixed stars appear to rise and set daily to the same bearings. Nature abounds in such regularities.

It is often said, then, that the scientist is not interested in unique events or phenomena. But this is only partly true. A unique phenomenon is the result of conditions that have not been repeated and cannot be reproduced by men. Yet some such events are of enormous scientific interest, for to explain them would vastly increase our understanding of the universe. How much would we not give to

125

know the origin of the universe, an absolutely unique event, or the origins of life and mankind, events which most biologists consider virtually unrepeatable.[4]

Consider some phenomena once thought unique. In November 1572 a new star appeared that outshone every other celestial body and amazed scholars all over Europe. In Sweden the young Tycho Brahe noticed it one evening as he walked home from his uncle's abbey, where he studied alchemy. Using his sextant, he calculated that it was outside the earth's atmosphere, beyond the moon, indeed among the fixed stars. Yet it was manifestly impermanent. As such it seemed to refute the classical assumption that the heavens were immutable. We now know that it was a supernova, the explosion of a star collapsing under the force of its own gravitation. (A supernova expels a shell of burning, gaseous matter that takes centuries to cool. In the Crab Nebula the telescope reveals the remnant of a far more spectacular explosion that was observed by the Chinese and Japanese in 1054.)[5] In October 1604 another supernova appeared over Europe and was observed by Brahe's former assistant, Johannes Kepler. This repeated challenge to the Aristotelian system excited Galileo, who six years later turned his telescope on the moon and found it pockmarked and imperfect.

Or take the coelacanth. This hulking, ugly, oily creature was netted by fishermen off Madasgascar in 1938 and then was spotted by the curator of the local museum. To her amazement it resembled the fish that had crawled out of the Devonian seas 370 million years ago to become the ancestors of 80 trillion land animals. Yet according to the fossil record these fish had been extinct for over 70 million years. She sent a sketch to the South African ichthyologist, J. L. B. Smith, who identified it. Of the specimen brought to his laboratory he said later, "I would hardly have been more surprised if I met a dinosaur on the street." The importance of the discovery lay not in the prospect of finding further specimens but in the recognition that this unprepossessing animal was one of the most crucial links in the evolutionary chain. As Michael Polanyi puts it, "Discoveries such as these are valued for the breadth of their implications, even though they establish no new general laws. They offer something more vague

126

and also more profound, a truer understanding of a large domain of experience."[6]

CLASSIFICATION

Unique events are scientifically important when they help us find order and regularity in the universe. But we cannot recognize phenomena as regular unless we classify them. Classification is the assignment of objects or events to groups according to their common properties. It depends on the fact that things resemble and differ from one another. But they do so in innumerable ways. Take the flowers in my garden. I could classify them by such properties as height, scent, color, month of flowering, or distance from my house. I must be careful then, to choose properties that define them in important ways. Hence I would probably ignore the property of distance. So, too, the chemist, classifying substances, picks out such properties as color, hardness, melting point, and reactivity with oxygen. He ignores such characteristics as the date when the substance was discovered or the location of the largest known sample.

What are the purposes of classification? They are to organize information, to aid the memory, and above all so to describe the structure and the relations of things that general statements can be made about them. What kind of classification serves the last purpose best? To answer this question, we must distinguish between natural and artificial classifications. A *natural* classification seeks to display the true natures of things. It groups phenomena according to what are thought to be their fundamental properties, in particular those that are related to other properties in important ways and hence can figure in generalizations leading to laws. An *artificial* classification divides phenomena according to readily observable, often superficial, characteristics. Hence it does not reflect the underlying natures of things and does not lead to important generalizations;[7] it enables us to identify the class to which something belongs—and nothing more. For example, human beings can be divided *naturally* according to their primary sex characteristics, which are correlated with a wide va-

127

riety of physical, physiological, and psychological traits, or they can be divided *artificially* into those weighing more than 100 pounds and those less, a trait which is related to few others.[8]

In the life sciences modern classification began in the late seventeenth century with the work of the English naturalist John Ray. Although Ray classified artificially, he was the first to describe species precisely. Moreover, he pointed the way toward a natural classification by categorizing species according to structures, such as the arrangement of teeth or toes in animals, rather than color or habitat. In the next century the Swedish botanist Carolus Linnaeus provided the binomial nomenclature and the hierarchy of class, order, genus, and species that have been used in biological classification ever since. He classified plants by the structure of their reproductive organs (the number and arrangement of their stamens and carpels), distinguishing between those plants with real flowers and seeds and those without them, and subdividing the former into bisexual and unisexual forms. Although he recognized that he had not achieved a natural system, he made biological classification a genuine discipline.

Later work brought biology closer to natural classifications. Early in the nineteenth century, for instance, Jean-Baptiste Lamarck separated spiders and crustaceans from insects as distinct classes, and distinguished between vertebrates with backbones (fishes, amphibians, reptiles, birds, mammals) and invertebrates without them. (Oddly enough, the invertebrates, defined by a feature they lack, make up 90 percent of all animal species!) Then Darwin's theory of evolution led to a much deeper classification of species. When biologists accepted the theory, they realized that similarity between species having many of the same characteristics was due not to natural affinity (as they had thought) but to evolutionary descent. More importantly, the linear-branching nature of evolution made it possible in principle to create a linear-branching classification system. The theory showed that, given enough information, living things could be arranged in a phylogenetic tree rather than in a set of discrete classes. Since then a linear-branching classification, to which successive subdivisions can easily be applied, has been worked out, reflecting the descent of species themselves.

Perhaps the most striking example of a natural classification is

Dmitri Mendeleev's periodic system of the elements. In 1869 Mendeleev proposed that if the elements are arranged in order of atomic weight, they fall into groups possessing similar chemical properties. Other chemists had observed an orderliness in the elements, but Mendeleev had the courage to declare that his principle was a fundamental law of nature, and that any apparent deficiencies in his table were the result of gross errors in the measurement of atomic weights, or else due to the fact that certain elements had yet to be discovered. In the next 16 years he was vindicated by the discovery of scandium, gallium, and germanium, whose existence he had predicted. Thirty years later, in an odd metamorphosis, the atomic number of an element, which simply listed its place in the table, was shown by Rutherford and Bohr to be the clue to its underlying structure. This number, they said, is identical with the number of protons in the nucleus, which is equal to the number of electrons surrounding it. Thus, an atom of uranium, then the heaviest known element, numbered 92 in the table, has a nucleus of 92 protons and has 92 electrons in orbit.

Today, with over 200 particles known and no accepted theory to explain them, particle physics faces a chaos like that of chemistry a century ago. In 1963, in a Mendeleevan move, Murray Gell-Mann and Yuval Ne'eman proposed that particles with the same spin and parity should be grouped together according to a mathematical scheme called SU (3).[9] The scheme predicted, for example, that there should be eight particles with "spin ½" and "positive parity," precisely the number that had been discovered. The scheme also predicted the charge, mass, spin, and parity of a missing particle, the omega minus. The experimentalists rose to the challenge. In a bubble chamber of the Brookhaven proton accelerator, then the largest in the world, 100,000 photographs were taken. These were scanned for weeks until at last, after 50,000 had been examined, the particle was spotted in a corner of one of them.

Unlike Mendeleev, Gell-Mann sought to explain the relations stated in his table. In 1966 he put forward his quark theory, still unconfirmed, which argued that there are three basic particles, called quarks, out of which all other particles are formed. Meantime, in 1974 a fourth kind of quark was added, bearing a proposed new prop-

129

erty of matter called "charm." As the quark theory was applied in detail during the next decade, it became clear that to explain the data, one must postulate yet another property that was capable of assuming three possible values. This property was nicknamed "color," though it had nothing to do with color in the ordinary sense. Added to each of four sorts of quarks, this property in effect tripled their number. Recent experiments suggest that there may be more quarks still. Thus the quark theory, introduced to explain the abundance of particles, may soon have to be explained itself.[10]

Classifying and generalizing often proceed together. If the scientist refuses to generalize until the phenomena have been classified, he cannot tell whether the classification proposed is likely to stimulate hypotheses. On the other hand, if he generalizes about phenomena that are inadequately classified, he cannot be sure that his hypotheses apply beyond his data. One should note, too, that to classify is itself to make a limited generalization that all members of a given class have such and such properties. To classify whales, for instance, as mammals rather than fishes is to generalize that all infant whales are given milk by their mothers.[11]

LAWS

The purpose of classification is to lead to laws—statements describing regularities.[12] A scientific law takes the form, "Whenever property A, then property B." A law may assert that whatever has A also has B— for example, that every bar of copper has a melting point of 1083 degrees Centigrade. This law describes a regularity of *coexistence*, a pattern in *things*. Or it may assert that whenever a thing having A stands in a certain relation to another thing of a certain kind, the latter has B—for example, that whenever a stone is thrown into water, it produces a series of expanding concentric ripples. This law describes a regularity of *succession*, a pattern in *events*. Classification schemes describe regularities of coexistence; laws describe regularities of both types.

A scientific law fulfills two main functions. First, it summarizes a great many facts and so makes for economy of thought, for if

130

we know the law, we need not remember the facts. Second, it enables us to predict further facts, for it tells us that if a phenomenon is an instance of the law, it will behave as the law states. Laws are of two types: empirical generalizations, sometimes called empirical or phenomenological laws; and laws of nature, sometimes called theoretical laws.

An *empirical generalization* is a statement made by inferring that what has been observed in particular instances of a phenomenon is observable in all instances. Some examples are: "Metals expand when heated," "Water seeks its own level," "Mammals manufacture their own vitamin C, and "Salt crystallizes in cubes." Many generalizations have become well-known laws, such as Kepler's three laws of planetary motion; Galileo's law of freely falling bodies; the law of the simple pendulum (that the period of a simple pendulum oscillating through a small arc is proportional to the square root of its length): the Boyle-Charles gas law (that the volume of a given mass of gas varies inversely with the pressure to which it is subjected); Ohm's law (that the electrical current flowing through a body is proportional to the applied voltage); and the second law of thermodynamics ("Heat cannot of itself pass from a colder to a hotter body"). Nevertheless, generalizations are not explanatory. The decreasing volume of a gas under pressure is not explained by referring to the generalization that in all cases the volume of a gas varies inversely with its pressure. To explain this phenomenon, as I shall show, we must use the kinetic theory of gases.

Empirical generalizations must be distinguished from *tendency statements*, which assert that a thing is apt to behave in a certain way under certain unspecified conditions. Biology is full of tendency statements, such as Cope's rule that body size tends to increase during the evolution of a group, Dacqué's "law" that contemporaneous species living in the same area tend to change in analogous ways, and Bergmann's principle that in colder regions the members of the geographical races of warm-blooded animals are larger than the members of the same species in warmer regions. These statements do not qualify as laws or generalizations because there are too many exceptions to them. For example, it has been found that 20–30 percent of Palaearctic and Nearctic birds and 30–40 percent of Palaearctic and

Nearctic animals are exceptions to Bergmann's principle.[13] Tendency statements often are proposed in the behavioral sciences where the conditions under which a putative regularity holds are unknown. Examples are Thorndike's "law" of effect, which asserts that in human learning an association tends to strengthen when followed by "satisfying" consequences, and Durkheim's "law" that suicides tend to increase in times of rapid economic change.

An empirical generalization becomes a law when it is incorporated in a theory. The law asserts the existence of a stable pattern in events or things, and the theory states the mechanism responsible for this pattern. The kinetic theory of gases explains the Boyle-Charles law by proposing the mechanism of a cloud of gas molecules colliding with one another; Newton's theory of gravitation explains Galileo's law of free fall by postulating a gravitational force of which it specifies the mode of operation but not the nature. The mechanism guarantees that the regularity is not accidental. For until we can point to a single underlying cause, we cannot be sure that the pattern is not the chance result of many different causes acting independently and so not a true regularity at all.[14]

Nevertheless, many generalizations have been called laws before being included in theories: for instance, Kepler's laws, Boyle's law, Charles's law, and the Wiedemann-Franz law of mutual variation between thermal and electrical conductivity. This usage reflects the greater confidence scientists place in empirical generalizations proper as opposed to taxonomic generalizations. A taxonomic generalization is regarded as simply a summary of evidence and hence refutable by a counterinstance. An empirical generalization, on the other hand, instead of being refuted by a counterinstance, normally is limited, modified, or idealized so that the apparent falsifier is made to fall outside its scope.

In some fields, irregularity in the behavior of individual phenomena is found to coexist with a regular trend in the behavior of a long series or large aggregate of such phenomena. In the long series or large aggregate, individual variations tend to cancel out, and statistical regularities emerge. These are described by *statistical laws*. Where universal laws have the form "Whenever A, then B," the latter have the form, "The probability that when A, then B, is *p*" or "The probability of an A being also B is *p*." Examples are the laws of quantum and statistical mechanics, of genetics, and of radioactive decay. Thus the Maxwell-Boltzmann law states that the average amount of energy used for each different direction of motion of an atom is the same. The law of radioactive decay states that all pure radioactive substances decay exponentially. For instance, 50 percent of the atoms in a piece of radium will disintegrate within 1,600 years. But the process cannot be predicted for the individual atom; we cannot say which of the atoms will disintegrate during this period but only that 50 percent of them will. We assert the probability that a given atom will decay by 50 percent within 1,600 years.

Other laws are formulated with the theories of which they are part, such as Maxwell's laws of the electromagnetic field, the laws of quantum mechanics, the principle of natural selection, and the Hardy-Weinberg law of population genetics.[15] These highly abstract laws are often called theoretical principles. They are the fundamental constituents of any science. A similar role is played by Newton's laws of motion and gravitation, though they are closer to the level of observation.

When laws become well confirmed, they are regarded as definitions of the terms they contain and hence as true by definition. They then are called *conventions*. Newton's laws of motion are often treated as conventions.[16] $F = ma$ is regarded both as a definition of force and as a rule specifying a technique for measuring forces. As such it is held to be unfalsifiable. The law of the conservation of energy, which began as an empirical generalization, is now a definition. Hooke's law (that in elastic bodies stress is proportional to strain) has become a definition of the elastic constant.

Laws and generalizations are made more accurate by being tested under diverse conditions. Such tests often reveal causal factors that were overlooked when the law was originally formulated. The melting-point of ice, for instance, once was declared to be 0 degrees Centigrade. Further investigation, however, revealed that under greater pressure ice tends to melt at a lower temperature. Hence the melting-point now is stated as a function of the pressure. Ordinary

133

hydrogen long was thought to be a pure substance. In 1931, however, it was found to contain a little deuterium, which can be isolated and has very different physical properties.

When a nontaxonomic law or generalization is confronted by counterinstances, the scientist can proceed in any one of three ways.[17] He can limit the scope of the law, expressing the counterinstances in the form of *boundary conditions*, which state the circumstances under which the law ceases to hold. He can modify the law by combining it with the boundary conditions to produce a new law. For example, the Boyle-Charles gas law, which states that $pv = RT$ (p = pressure, v = volume, R = the gas constant, and T = temperature), was found not to hold at high pressures and low temperatures. These boundary conditions were then incorporated by Johannes D. Van der Waals into the equation bearing his name,

$$(p + \frac{a}{v^2})\, (v - b) = RT$$

where a/v^2 is a correction for the mutual attraction holding between the molecules and b is a correction for the actual volume of the molecules themselves. Finally, the scientist can idealize the law, stating that it holds only for ideal entities lacking those properties that would compel him to state boundary conditions for the law as originally formulated. Thus the Boyle-Charles gas law applies to the behavior of ideal gases, and the laws of mechanics apply to the motions of ideal bodies. Both laws have to be corrected when they are used to predict the behavior of actual phenomena.[18]

Laws as Mathematical Equations. When the facts described by the law are expressed numerically (that is, when numbers are used to represent the amounts by which the properties mentioned in the law covary), the law can be written as a mathematical statement. As in the case of the melting of ice under pressure, the scientist often wishes to find a "functional relation" between changing properties. He may, for instance, wish to determine how the length of an iron bar changes with the temperature of the iron, how the volume of a gas changes with the pressure exerted on it, or how the distance covered by a body falling freely near the earth's surface changes with time. A *function* is a relation between two varying quantities or variables, which specifies that a given value of one quantity is accompanied by a corresponding value of the other. It is expressed symbolically as $y = f(x)$, meaning that y is a function of x, or y varies as x does.

Suppose the scientist wishes to find a functional relation between electric current and voltage in a conductor. He begins by altering the intensity of the current by fixed amounts and noting the corresponding voltages. He may obtain the following figures:

Current (C)	Voltage (V)
1	2
2	4
3	6
4	8
5	10

(C and V stand for any numbers the scientist may obtain by measuring the intensity and voltage of the current.) He then looks for an exact relation between the two sets of numbers which he can express as an algebraic function. Often he does not find one. He may discover, for instance, that the numbers vary independently of one another, which signifies that the properties represented by the numbers are not regularly correlated. In our case, however, he discovers that the number representing the current is always twice that representing the voltage. He expresses this relation in the function $V = 2C$.

To check this result, the scientist carries out another series of experiments using a different conductor and finds that this time the current intensity is always three times the voltage, a relation he expresses in the function $V = 3C$. He now can make a further generalization. Just as the numbers obtained by measurement can be represented by the number variables V and C, so the relation between these numbers, which is designated by the numerical constants 2 and 3, can be represented by a number variable standing for the numerical values of a third property of the phenomenon under investigation. This property is responsible for the voltage being twice the current intensity in the first series of experiments and three times that intensity in the second series. The scientist guesses that is is the resistance of the conductor, and he symbolizes it by the letter R. Using R as a variable, he writes the function, $V = CR$, which expresses a universal generalization derived from the results of his experiments. This equation summarizes his observations and can be used to predict all future observations of this class of phenomena. It states that the magnitude of the voltage is always equal to the product of the intensity of the current and the resistance of the conductor. If this functional generalization is widely confirmed by the experiments of other scientists, it may be called a law, and if it is incorporated into a theory, it will acquire not only the title of a law but also the status of one. (It is in fact Ohm's law, and it is explained by Maxwell's electromagnetic theory.)

Abstraction and Idealization. Let me mention in passing that many phenomena are too complex to be understood in complete detail. In physics and chemistry, though less so in biology, the scientist often ignores many features of a phenomenon and instead abstracts a few properties that together form a simplified version of it. In studying moving bodies, for instance, classical mechanics considers only such properties as mass, velocity, and distance traveled over time, and ignores such properties as color and texture.

The scientist also stipulates that these properties hold under idealized conditions. Velocity, for example, is treated as speed in a frictionless environment, and mass as the substance an object would have if it were concentrated at a dimensionless point. An object falling from a table is represented as having fallen in a vacuum, been a mass-point, and so on. The ideal concepts of physics include perfectly rigid bodies, incompressible fluids, and frictionless planes. In chemistry there are pure compounds, adiabatic processes (processes occurring without a change in heat), and infinitely dilute solutions. Thermodynamics, statistical mechanics, and quantum mechanics deal with similar idealizations. Boyle's and Charles's gas laws describe the behavior of ideal gases, not real ones. The valence theory of chemistry treats reactions between theoretically pure chemical substances, ideal fictions that actual substances only approximate. The modern, synthetic theory of evolution explains evolutionary phenomena as the result of changes in the distribution of genotypes (genetic makeups of individuals or groups) in a population. It treats populations of individuals as idealized populations of genotypes whose distribution alters in response to changes in only a few chosen factors, such as reproductive rates, reproductive barriers, and crossover frequences. Scientists do not claim that these idealizations actually exist, but only that real phenomena approximate their behavior within our ability to measure them.

THE STRUCTURE OF THEORIES

The most important constructs of science are its theories. A theory explains a law by proposing a mechanism that accounts for the regularity described by the law and by entailing the law as a logical consequence of its assumptions. Take as an example the kinetic theory of gases.

The purpose of this theory is to explain the behavior of gases under a range of conditions. Before the theory was formulated, various generalizations about the behavior of gases already had been proposed and confirmed. It then was seen that these generalizations could be connected logically, and the regularities they expressed could be explained, if it was assumed that a gas is really a swarm of minute, colliding particles moving at high speeds in straight lines. From these assumptions, together with Newton's laws of motion, certain known generalizations about the behavior of gases were deduced, including Boyle's and Charles's laws describing how gases react to varying temperatures and pressures. But scientists also made various deductions about some hitherto unsuspected properties of gases, such as viscosity, diffusion, and heat conduction. Like all good theories, then, the kinetic theory not only organized and explained a number of known laws but also yielded further testable generalizations.

What is the structure of this theory? And do all theories have the same structure? Let us examine two recent attempts to articulate a structure common to all scientific theories.

137

Logical Empiricism: Theories as Calculi. According to the logical empiricists, a scientific theory ideally can be given a tripartite structure consisting of a calculus, a set of correspondence rules, and a model. A *calculus* is a deductive system of axioms and theorems, written entirely in logical symbols, that refer to nothing in the outside world. Its purpose is simply to display the inner logical structure of the theory as clearly as possible.

The calculus is related to the empirical content of the theory through *correspondence rules*, sentences correlating certain logical terms in the calculus with observation terms describing the phenomena the theory is intended to explain. For example, the temperature of a gas is assumed to be proportional to the average kinetic energy of the gas molecules. A correspondence rule expressing this equality coordinates the observation term "temperature" with the term in the calculus to be designated "kinetic energy."

Finally, the empirical content of the theory is represented by means of a set of sentences called a *model*. The sentences are obtained by substituting for the uninterpreted terms of the calculus other terms with which we are already familiar, such as volume, temperature, and pressure. Some of these familiar terms appear in known laws (e.g., Boyle's and Charles's laws) having the same logical form as some statements of the calculus. Others, in the case of the kinetic theory, are taken from classical mechanics—for example, molecule, velocity, mass. For most formulae in the calculus we can write equivalent sentences in the model referring either to properties or to observable effects of the gas molecules. The model is in this sense an interpretation of the calculus. For the logical empiricists, however, the essence of the theory remains the calculus. The model, says Ernest Nagel, is only a "heuristic aid . . . serving as a guide for setting up the fundamental assumptions of the theory, as well as a source of suggestions for extending the theory."[19]

What do we gain by reconstructing a theory this way? Undoubtedly we display more clearly the deductive structure of the theory (insofar as the theory has one), for we distinguish, for instance, between assumptions and theorems, and we give each term in the theory a precise meaning. Nevertheless, the gain in clarity is more than offset by the loss in relevance, for the calculus says nothing whatever about the content of the theory. This content, including the statement of the explanatory mechanism, is what interests the scientist who proposes the theory and the scientists who use it. It is more reasonable, and more in accord with practice, to regard the model as the essence of the theory and the calculus as a refinement. Again, as Carl G. Hempel points out, a theory may be axiomatized in many ways; the most economical or elegant of these, which the philosopher may choose for his reconstruction, need not correspond to the way in which the theory actually is used. For example, if Newtonian mechanics is axiomatized, the second law of motion can be presented as a definition, a premise, or a theorem. Yet the role played by the law in the axiomatized system does not tell us whether scientists themselves use the law as a definitional truth, a theoretical principle, or a derivation.[20]

Harré on Theories. Let us then regard the model as the heart of the theory and the deductive organization as an optional heuristic aid. This is how Oxford philosopher Romano M. Harré reconstructs a scientific theory. According to Harré, the fundamental activity of science is the search for and elaboration of models. A theory, he says, is essentially a model of a currently unknown mechanism in nature.[21] Darwin's theory of evolution, for instance, is essentially the model of natural selection. The scientist creates his model by analogy with a model already in use. Thus a model has both a source and a subject. The source is whatever the model is based upon (e.g., an analogous model); the subject, whatever the model explains. The sources of Darwin's model are Malthus's theory of population growth and the selection of favorable varieties by domestic plant and animal breeders. The subject is the multiplication and variation of species in nature.

If the use of a model enables scientists to propose and confirm a range of hypotheses, then, says Harré, the mechanism described by the model comes to be regarded as real (something that *does* exist) and no longer hypothetical (something that *may* exist). Descartes, for instance, considered the heart as a furnace, whereas Harvey thought of it as a pump. Descartes was wrong. The heart *is* a pump, because it is the pumping action of the heart that makes the blood circulate. By proposing models, then, and investigating the hypotheses they yield, scientists discover how nature really works.

Harré divides a scientific theory into three parts: model, transformation rules, and empirical laws.[22] Transformation rules relate statements in the model to statements about observable phenomena described by the laws. The kinetic theory of gases, for example, consists of: a *model*, including such statements as "molecules exist" and "collisions are elastic"; *transformation rules*, such as "pressure is caused by molecular impacts" and "temperature is the average kinetic energy of the molecules"; and *empirical laws* such as $pv = RT$.[23]

MODELS

I have spoken of models as parts of theories, but the term "model" is used more widely than this; indeed, it is one of the most overworked terms in science. Fortunately for us, however, most things called models can be classified as either representational, theoretical, or imaginary.[24]

A *representational model* is a three-dimensional physical representation of something, such as a museum model of the solar system, an engineering model of a dam or airplane, or a colored-ball model of the structure of a molecule. A variant is the analogue model, which represents an object without reproducing its properties, as in the case of an electrical circuit used as a model of an acoustical system.

A *theoretical model* is a set of assumptions about an object or system. (A system, unlike a particle, is an object with parts.) Examples are the billard-ball (spherical particle) model of a gas (first proposed by Scottish physicist John James Waterston, a keen billiards player!), the corpuscular model of light (according to which light consists of moving particles), and the Watson-Crick helical model of the DNA molecule. A theoretical model may be expressed in the form of mathematical equations, but it must be distinguished from any diagrams, pictures, or physical constructions used to illustrate it. Thus the Watson-Crick theoretical model is distinct from the representational models the two scientists built on their way to achieve it. A theoretical model attributes to the object or system it describes an inner structure or mechanism that is responsible for certain properties of that object or system. For example, the corpuscular model of light attributes a particle structure to light in order to explain such properties of light as reflection and refraction. The properties explained by the model may be macroscopic, as with the gas model, or microscopic, as in the case of Bohr's atomic model. The mechanism or structure proposed by the model may also be either microscopic, as in the atomic or gas models, or macroscopic, as in the astronomical model of the origin of the universe. Theoretical models are the most important type of model used in science. Many, such as those mentioned here, are regarded as theories in their own right, and they are described as such in this book.

An *imaginary model* is a set of assumptions proposed, not as a plausible description of an object or system, but as a description of what the object or system would be like if certain conditions were satisfied. For example, Henri Poincaré postulated an imaginary world governed by the axioms of Lobachevsky's non Euclidean geometry and described how it would appear to an inhabitant. Maxwell's mechanical model of the electromagnetic field is imaginary in this sense. Instead of claiming that the electromagnetic field is in fact governed by the laws of Newtonian mechanics, Maxwell described what it would be like if it were governed by them. An imaginary model may serve other purposes. It may show that certain assumptions, thought to be contradictory, are at least consistent (e.g., that a fully mechanical electromagnetic field is logically possible). It may lead to further investigation of an object or system on the supposition that the imaginary structure it proposes may be like, or may illuminate, the real one. Or it may improve our understanding of the assumptions it consists of by providing an application of them.

MATHEMATICS AND THEORY-BUILDING

Theories and models often are constructed and expressed mathematically. How is this done? Basically, mathematics provides the scientist

141

with a range of deductive structures by means of which he can infer the implications of statements—such as empirical laws or theoretical principles—that are isomorphic with, or have the same logical form as, propositions in the mathematical structures themselves.

A mathematical structure consists of a set of axioms and a set of theorems that are logically deduced from them. Both axioms and theorems set forth the general relations that hold between purely abstract entities. The scientist interprets this structure by replacing the symbols or variables in certain axioms or theorems with subject-matter terms of his own. So interpreted, the abstract mathematical propositions become statements about the world.

Mathematics is used to build models and theories in three main ways. The first, and least common, way is to construct a mathematical formalism and then to interpret it physically. This was how Erwin Schrödinger developed his theory of wave mechanics from an earlier theory proposed by Maurice de Broglie. In his dissertation (1924) de Broglie put forward a beautiful set of equations expressing the idea that matter, like radiation, has both wave and particle properties. Persuaded by the beauty of these equations that de Broglie's physical idea must be correct, Schrödinger sought to modify the principal equation so that it would apply to electrons within the atom. In 1926 he hit upon an equation which he considered so beautiful and, in essence, so simple that it was very likely to be true. Schrödinger, then, sought to capture the truth of nature by seeking beauty in his equations. Nevertheless, he also was guided by a physical idea, for he believed that at the microscopic level matter behaved in wave as well as particle fashion.

More often the scientist begins with a physical idea and then seeks to make it more precise by expressing it mathematically. Maxwell, for example, made Faraday's theory of the electromagnetic field more precise by casting it in the form of differential equations, and also more concrete by representing it in mechanical models described by these equations. In creating his theory of general relativity, Einstein arrived at the idea that gravitation is related to the geometrical structure of space, but he lacked the mathematical knowledge to represent this structure clearly. So in 1912 he turned to a mathematician friend, Marcel Grossman, and asked him to put his idea in mathematical form. Grossman did so by using Bernhard Riemann's four-dimensional geometry of curved spaces, treating time as a fourth dimension.

Finally, the scientist uses mathematics to deduce the consequences of his assumptions. Maxwell, for instance, deduced that electromagnetic radiation travels at the same speed as light, and hence that light must be a form of electromagnetic radiation. Similarly, Paul Dirac united quantum theory with special relativity in a set of equations, from which he deduced, among other things, that there are electrons of positive charge. Although such particles were thought at the time to be impossible, Dirac insisted that they must exist since they were the logical consequence of assumptions that he and other physicists had every reason to believe were true. Five years later he was vindicated when Carl Anderson discovered experimental evidence of these particles.

TELEOLOGY

So far I have discussed certain intellectual constructs common to all the sciences. Now I turn to a question specific to the life sciences: Are there any goals in nature, and, if so, how can they be explained? Everywhere in the living world we seem to see goal-directedness. The organs and behavior of plants and animals are adapted to particular ends. The eye is adapted to seeing, the hand to grasping, the wing to flying. How are such phenomena to be accounted for?

In science a particular phenomenon often is said to be explained when a statement of its occurrence is deducible from one or more laws or generalizations together with statements of antecedent conditions. For instance, the fact that a particular piece of iron becomes demagnetized on heating is explained by deducing a statement of this fact from the generalization, "Magnetism ceases above the Curie temperature (770 degrees Centigrade for iron)," together with statements describing the nature and intensity of the heat source. The force of the explanation lies in showing that this case of demagnetization is not unique but an instance of a general regularity. Many biological phenomena are explained like this. The flight of an insect,

143

for example, is explained in terms of the laws of aerodynamics. Similarly, a heart attack can be explained by reference to the law that chemical deposits from ingested foodstuffs lead to arterial occlusion.

Adaptive phenomena, however, are explained by specifying the goal or function they serve. Why, for instance, does a fish have gills? To provide a mechanism for breathing. Why is a squirrel carrying straw? To build a nest. Such explanations are called "teleological" after the Greek word *telos* meaning end or goal. They explain phenomena by reference not to their causes but to their supposed consequences. More precisely, a teleological account explains the presence or occurrence of some property in an organism by showing that it contributes to the existence of some other property of the organism. Thus sweating and shivering are explained teleologically by the supposition that they keep the organism's temperature constant.

Teleological explanations are used to account for three main types of phenomena: (1) Purposive behavior, that is, behavior whose goal or end-state is consciously entertained by the agent. Human beings generally behave purposively, and animals appear to. A rabbit sniffing the air, a dog barking, a bird migrating seem to be purposive. (Whether they actually are is a question that scientists and philosophers have yet to settle.) (2) Homeostasis, in which a mechanism enables an organism to achieve or maintain a certain property notwithstanding changes in the environment and in the organism itself. Examples are the elimination of waste from the blood by the kidneys, and the maintenance of a constant body temperature by sweating and shivering. (3) Organs that are designed to perform a certain function, as the ear is made for hearing and the nose for breathing.

The use of teleological explanations is justified by the theory of evolution. One may reasonably ask what an organ, homeostatic mechanism, or pattern of behavior is for, since it hardly would have survived if it did not contribute to the reproductive fitness of the organism. An organism has certain organs, homeostatic mechanisms, and patterns of behavior because it has descended from organisms whose ability to produce fertile offspring depended on possessing these characteristics. A characteristic that increases the reproductive fitness of an organism will be selectively favored and in time will spread throughout the population. Take the regulation of body temperature.

144

A man dies if his temperature rises or falls more than a few degrees above or below normal. Hence natural selection has favored the homeostatic mechanisms of shivering and sweating. In cool weather shivering generates heat, and in warm weather sweating dissipates it.

Almost every characteristic of an animal or plant is teleological in that it contributes *proximately* to the attainment of a particular goal or end-state and *ultimately* to reproductive efficiency. Natural selection itself is proximately teleological in that it produces and maintains organs, homeostatic mechanisms, and behavior patterns whose consequences contribute to the reproductive efficiency of organisms. But natural evolution is not ultimately teleological. It does not tend to produce organisms of any particular kind, and the process of evolution has no final goal. Natural selection occurs contingently in response to changes in a species' environment.

Nevertheless, can teleological explanations be reformulated, without loss of content, as causal ones? A number of philosophers have argued that they can. Ernest Nagel, for example, has proposed that teleological statements really are telescoped causal arguments, which can be fully articulated by treating functions and goals as necessary conditions.[25]

Take a typical teleological statement: "The function of chlorophyll in plants is to enable them to perform photosynthesis, i. e., to form starch from carbon dioxide and water in the presence of sunlight." Such a statement accounts for the presence of a certain feature A (chlorophyll) in every member of a class of systems S (plants) which possess a certain organization C of parts and processes. It does so by asserting that when S is placed in a certain environment E (when it is provided with water, carbon dioxide, and sunlight), it will perform a function F (photosynthesis) only if it has A (chlorophyll). This statement, says Nagel, is a condensed argument whose content can be expressed as follows: When supplied with water, carbon dioxide, and sunlight, plants manufacture starch; if they lack chlorophyll, however, they do not manufacture starch, even when supplied with water, carbon dioxide, and sunlight. More generally, a teleological statement of the form, "The function of A in system S with organization C is to enable S in environment E to engage in process P," can be unpacked into the nonteleological argument: "Every system S with

145

organization C and in environment E engages in process P; if S with organization C and in environment E lacks A, then S cannot engage in P; hence S must have A." The presence of chlorophyll is explained, then, by saying that it is a necessary condition for the performance of photosynthesis, which in turn is necessary if the plant is to maintain itself.

On this analysis, teleological explanations are logically equivalent to causal ones. Both types of explanation have the same content; they differ only in the emphasis they give to it. A teleological explanation directs our attention to the "*consequences* for a given biological system of a constituent part or process." The equivalent nonteleological explanation points to "some of the *conditions* . . . under which the system persists in its characteristic organization and activities." The difference is "comparable to the difference between saying that Y is an effect of X, and saying that X is a cause or condition of Y." [26] Whether we assert that photosynthesis is an effect of the presence of chlorophyll, or that the presence of chlorophyll is a cause or condition of photosynthesis, we say the same thing.

Or do we? A nonteleological explanation only asserts that the presence of chlorophyll is a cause of photosynthesis; it does not explain why the chlorophyll is there. The connection between chlorophyll and photosynthesis remains contingent. Chlorophyll happens to be present in plants and happens to make photosynthesis possible. A teleological explanation, on the other hand, implies that chlorophyll is present in plants precisely in order to make photosynthesis possible. As philosopher Francisco Ayala puts it, "Teleological explanations imply that the end result is the explanatory reason for the *existence* of the object or process which serves or leads to it." [27]

Teleological explanations, then, are essential to biology. They imply that the parts, processes, and behavior patterns of living things are organized so as to attain specific goals, which contribute as a rule to the ultimate goal of reproductive fitness. The question "What are these structures and processes for?" can only be answered in teleological language. Teleological statements cannot be reformulated in causal terms without sacrificing this essential implication. [28]

146

REDUCTIONISM

Reductionism is another issue that especially concerns biology. As I have explained, a theory is said to be "reducible" to another when it can be deduced from the other as a logical consequence or special case. The reduced theory T' is held to be explained by the reducing theory T in the sense that (a) it is entailed by T and (b) the patterns it attributes to nature are accounted for by the causal mechanism proposed by T. Ernest Nagel distinguishes between two types of reduction: "homogeneous reduction," when the two theories share the same terms, as in the alleged reduction of Galileo's law of freely falling bodies and Kepler's laws of planetary motion to Newton's theory of mechanics and gravitation; and "inhomogeneous reduction," when the theories contain different terms, as in the alleged reduction of thermodynamics to statistical mechanics and of optics to electromagnetism.[29] In cases of inhomogeneous reduction, the different terms are correlated by means of correspondence rules. Thermodynamics treats the behavior of macroscopic objects such as gases, whereas statistical mechanics deals with the behavior of microscopic entities such as gas molecules. In order to reduce the one theory to the other, macroscopic concepts, such as the concept of the temperature of an ideal gas, must be identified with microscopic ones, such as the concept of the average kinetic energy of the gas molecules.

But is Nagel correct? In the history of science there probably has been one entirely successful reduction—Maxwell's reduction of the wave theory of light to his electromagnetic theory, which identifies light waves with electromagnetic waves.[30] In most cases of alleged reduction (such as that of Newtonian mechanics to special relativity and quantum mechanics), so many terms change meaning between theories that few terms with shared meanings can be used in correspondence rules; as a result, strict logical deduction is impossible.[31]

As a research *strategy*, however, reduction has the merit of leading us to propose new, more comprehensive theories to unify existing knowledge and open fresh fields for investigation. Newtonian mechanics, relativity, quantum mechanics, and the theory of the structure of DNA all followed from attempts to unite accepted theories with broader ones. In the physical sciences especially, the search for reductions serves as a powerful stimulus to theory-construction. It acts as a brake only when (as the logical empiricists advocated) it is used as a criterion for rejecting new theories that do not strictly entail the established theories.[32] How useful is this strategy in the life sciences?

In biology three main forms of reductionism may be distinguished: ontological, epistemological, and methodological.[33]

Ontological reductionism is the thesis that the atomic and molecular processes underlying all living phenomena can be explained fully in physicochemical terms. This thesis is opposed to the "vitalist" contention that living matter is infused by a nonmaterial principle—a "vital force" or *"élan vital,"* which is responsible for the organic characteristics of growth, adaptation, and reproduction. Ontological reductionism, however, does not exclude the possibility that above the atomic and molecular levels living phenomena may be subject to biological rather than physiochemical laws. As Theodosius Dobzhansky has said, "Most biologists . . . are reductionists to the extent that we see life as a highly complex, highly special, and highly improbable pattern of physical and chemical processes."[34] The key word here is "pattern." Most biologists hold that although life is based on inanimate matter, it possesses properties that do not belong to its separate inanimate constituents but only emerge when these constituents are arranged in certain ways. These biological properties are peculiar to whole entities—to the cell, the organ, or the organism—and can be discovered only through the study of those entities. This point of view sometimes has been called "emergentism" or "organicism."

Epistemological reductionism is the thesis that all biological laws and theories in principle can be derived from physicochemical theories about atomic and molecular processes. It is generally agreed, however, that no such reduction has yet been carried out. For instance, no one has succeeded in reducing Mendelian to molecular genetics. There are various reasons for this. In Mendelian genetics the gene transmits traits from one generation to the next; in molecular genetics it not only transmits traits but also develops them (through transcription into RNA molecules and translation into proteins). Again, Mendelian terms such as "dominant" and "recessive" cannot be correlated consistently with molecular mechanisms. It sometimes turns out, for instance, that the same molecular state of affairs leads to phenomena that would have to be described by using two or more different Mendelian terms.[35]

The strategy of theory-reduction can lead to the proposal of more comprehensive theories, but it need not imply reduction of biological theories to physicochemical ones.[36] Actually, only a few biologists today insist that biology either is or is not reducible to physics and chemistry.[37] Most maintain that this is an empirical question to be answered by continued investigation.

Methodological reductionism is the thesis that solutions to biological problems should be sought mainly at lower levels of complexity and ultimately at the atomic and molecular levels. Most biologists agree, however, that living phenomena should be studied at all levels of organization and that higher-level phenomena generally require explanation in teleological terms. Recent research in antibody formation, for instance, after much analysis of antibodies at the molecular level eventually turned to the study of cells, organisms, and populations.[38]

SCIENCE AND OTHER KINDS OF KNOWLEDGE

We have examined the internal structure of scientific knowledge—its hierarchy of data, classification schemes, laws, and theories. Now let us look at it externally. What is the scope of this knowledge? How much does it tell us about the world? What is its relation to other forms of knowledge?

Science is only one part of mankind's attempt to understand the world in all its aspects. Man strives to discover an order in the flux of experience, whether this order be observed, as in the recurrence of the seasons, or whether it be postulated by sophisticated theories such as those of relativity, quantum mechanics, and evolution. The quest for order in experience unites science, literature, history, religion, philosophy, and the arts. Science seeks this order in

149

man's experience of nature; literature and the arts seek it in man's inner experience and in his relations with his fellows; history, in the human past; religion, in man's relation to a Supreme Being; philosophy, in all these enterprises themselves.

Science both narrows and adds to the experience of nature. It narrows that experience by striving to eliminate everything in it that is purely personal. It seeks to remove whatever is unique to the individual scientist, such as the memories, emotions, and aesthetic feelings stirred by the arrangements of atoms, or the colors and habits of birds, or the expanse of the Milky Way. It also endeavors to banish whatever all people experience but in different degrees depending on the perspective and the physical conditions of the experience. Hence science strips away much of the sensuous appearance of nature. Sunsets and waterfalls are described in terms of frequencies of light rays, coefficients of refraction, and gravitational or hydrodynamic forces. Clearly, this account, however enlightening, is not a complete explanation of what we actually experience.

In striving for objectivity, science excludes all reference to subjective experience, individual or communal. Hence, on the face of it, science describes a world of valueless things interacting as if mankind did not exist. But since the nature we experience is permeated by our evaluations—as in the terror of tornadoes, the calm of pools, and the sweet sadness of falling leaves—the scientific description of nature remains cold, incomplete, unsatisfying.

On the other hand, science adds to our knowledge by correcting our immediate experience of nature. Science does not replace such experience but transcends it, for immediate experience is our first and most biased encounter with nature, and it frequently is wrong. In immediate experience we meet solid objects and colors, but science has shown that a solid object really is a swarm of particles and that its colors do not inhere in it. Science begins precisely because we cannot adequately understand or control nature within the limits of ordinary experience.

But there is more to science than abstraction. Biology provides us with a wealth of information about living things that adds greatly to our joy in them. And physical science enhances our appreciation of nature by explaining phenomena of which we otherwise would know little. When we walk through a forest, we know that the wild-life around us has evolved over many centuries from much more primitive forms. We know that energy is stored in the standing boulder and expended in the cascade. We know that living things are made of cells and inert objects of atoms, that light is an electromagnetic field traveling in waves, and heat the motion of molecules. Knowing all this, does not nature mean more to us?

Again, if science disregards the beauty of nature's appearance, it does so in order to seek an intellectual beauty in nature's structure. Just as the beauty and emotional power of a work of art belong principally to the form with which the artist has unified a range of materials, so the beauty of a scientific theory lies in the simplicity of structure with which it describes an ultimate harmony in nature itself. Scientists seek those theories in which the fewest premises yield the most conclusions. Such theories have great aesthetic appeal, since they reduce the complex to the simple. When the scientist deduces known laws from deeper assumptions, he shows that the countless phenomena of nature behave as they do and fill the senses with delight, because they are the manifestations of an underlying structure in the world that is strange, fascinating, and simple.

Science, then, disvalues neither man nor his world. It illuminates the beauty in nature's order and expresses that beauty in the order of its own theories. The scientist studies nature not simply because it is useful but because he delights in it. He sees beauty in the harmony of nature's parts which his mind is able to grasp. "Intellectual beauty," wrote Henri Poincaré, "is sufficient unto itself, and it is for its sake, more perhaps than the future good of humanity, that the scientist devotes himself to long and difficult labors."[39]

Literature and the arts seek to portray aspects of human life both inward and social. Where they depict natural phenomena and take abstract forms, they do so as a rule to convey human qualities and possibilities in nonhuman things. A novel or a poem endeavors to present some aspect of human life with an order, scope, and intensity unattained in actuality. It aims to move and enlighten the reader with its vision of a significance in life that is glimpsed only fleetingly or not at all in the daily round. A work of art or literature abstracts from actuality in order to re-present it with greater richness and clarity. It stirs both the intellect and the feelings and so integrates the psyche.

History narrates what men have done and suffered in the past. Like science, it deals with fact, not fiction. But unlike science, and unlike art and literature, history chronicles sequences of events; it acquaints us with the collective transgenerational character of human life. Properly narrated, the history of science not only records the facts of scientific discovery but also displays in them the working of human nature, intellectual, emotional, and social.

Through religion man relates himself to a Supreme Being. This relation is studied by theology, celebrated by ritual, and enacted in prayer and the moral life. The truths to which religion lays claim concern the nature of God, the nature of good and evil, man's capacity for fulfillment or self-destruction, his relations with his fellows, his moral code, and his destiny in an afterlife. Unlike science, religion rests on revelation and seeks to sanctify the believer. Religious beliefs cannot ultimately be tested empirically but only by an inner experience of certainty, the experience of faith. Religion may indeed at times compete with science. More often, it complements science, since it serves different functions and possesses its own knowledge about other spheres. As we shall see, many scientists have been and are deeply religious, inspired by the belief that ultimately nature must reflect the mind of its creator.

In mystical experiences, which may be ecstatic or meditative, the individual is said to reach a state of oneness with the universe and to feel a deep peace. Science can explain the behavior of the nervous system during these experiences, but it cannot validate the experiences themselves, since they are unique to the individual and often purport to include contact with the supernatural. Mysticism, practiced in all cultures, claims to provide a unique, personal insight into reality unattainable by other means. Its claims should be treated with respect, if not always accepted.

Science, literature, art, history, religion, and mysticism illuminate aspects of reality. Philosophy endeavors to see reality whole. It analyzes the nature and findings of the different branches of knowledge, examines the assumptions on which they rest and the problems

152

to which they give rise, and seeks to establish a coherent view of the whole domain of experience.

Each of these forms of knowledge deserves to be cultivated for itself. Each in its own way acquaints us with some part of reality. We must see science in its place and not expect it to assimilate or discredit these other endeavors.

SUMMARY

Science seeks to discover relations in nature at different levels of complexity. To this end, scientific research consists in observing the course of nature, noting its regularities, and attempting to discover the unobserved causes or "mechanisms" from which these regularities spring. Regularities are persisting or recurring patterns among phenomena. In order to discover regularities, we must recognize similarities among phenomena, and to do this we must classify phenomena. To classify is to group phenomena according to common properties. Classifications may be artificial or natural. An artificial classification enables us simply to locate a phenomenon in a class. A natural classification enables us to relate it to other phenomena in terms of fundamental properties.

Natural classification is intended to lead to laws and generalizations—statements summarizing regularities. A generalization is a statement reached by inferring that what is true of certain observed cases is true of all of them. It differs from a tendency statement, which asserts that what is true of the observed cases is true of most of them. Generalizations and tendency statements differ in turn from statistical laws, which state that what is true of the observed cases is true of a certain percentage of all cases.

As a rule a generalization becomes a law when it is incorporated into a theory. The theory specifies the underlying cause, or mechanism, thought to be responsible for the regularity described by the law. The existence of a mechanism guarantees that the regularity is the result of a single causal process rather than the chance effect of many such processes operating independently. It thus guarantees that

153

the regularity is genuine rather than apparent, and that it is part of the real order of nature.

Some laws and generalizations are expressed numerically. A functional law or generalization, for instance, states the amount by which one quantity varies with respect to another. However, quantities (derived from measurable properties) are abstracted from the objects in question, and together represent simplified versions of them. These properties also may be represented as holding under idealized conditions. Thus the physical sciences in particular tend to deal with highly rarefied versions of actual phenomena.

Scientific knowledge, then, consists of empirical knowledge—data, classification schemes, generalizations, and laws describing patterns among things and events—and theoretical knowledge of the mechanisms or causes that produce these patterns. In a nutshell, science seeks to describe the things and events of the physical universe by classifying them and by expressing their interrelations in laws and generalizations, and it seeks to explain these laws by unifying them in theories.

Two major attempts recently have been made to display the structure of scientific theories. According to the logical empiricists, a theory can be reconstructed as a logical calculus, a set of correspondence rules, and an empirical interpretation or model. However, this account undervalues the empirical content of a theory and tends to ignore the version of the theory in actual use. Harré's reconstruction is more illuminating. For Harré, a theory is basically a model accompanied by transformation rules and empirical laws. His account emphasizes the creation of scientific theories rather than their formalization later.

Models are widely used in science. Three main types may be distinguished: representational, theoretical, imaginary. The first is a physical representation of an object; the second, a set of assumptions about an object or domain; the third, a set of assumptions about what an object or domain would be like under certain conditions.

Mathematics is used by theorists in three main ways. Sometimes, like Schrödinger, the scientist creates a mathematical formalism and then interprets it. More often, like Maxwell and Einstein,

154

he turns to mathematics to express a physical hypothesis more precisely. In both cases, however, he employs mathematics to deduce the testable consequences of his assumptions.

The life sciences use an intellectual construct of their own—the teleological explanation. Adaptive phenomena—purposive (and seemingly purposive) behavior, homeostasis, organs—can only be explained by identifying the goal or function (*telos*) they serve. This goal in the first instance is specific to the organism, but ultimately the goal is reproductive fitness and so the survival of the species. Hence teleological explanations are justified by the theory of evolution. They cannot be reformulated as ordinary causal explanations, which ignore the intrinsic connection between the adaptive phenomenon and its consequences.

In the physical and the life sciences researchers have sought to reduce certain theories to others. Only one complete successful reduction has been achieved—Maxwell's. As a research strategy, however, reduction has paid off by stimulating the invention of more comprehensive theories. Most biologists hold that the atomic and molecular processes involved in life ultimately can be explained by physicochemical theories (ontological reductionism). Whether biological theories about life at higher levels of organization ultimately can be reduced to physicochemical theories (epistemological reductionism) generally is regarded as an open question. Most biologists, however, are opposed to methodological reductionism; they hold that organic phenomena should be studied at all levels rather than at lower levels alone.

Science narrows the experience of nature by eliminating the purely personal response and thus the immediate experience of nature's sensuous beauty. But science adds to that experience by vastly increasing our knowledge of nature and correcting mistaken beliefs about it. Science also reveals the beauty of nature's underlying order which cannot be perceived by the senses. Nature, however, is only part of reality. To understand the other parts, we must turn to other intellectual quests: literature, history, religion, philosophy, and the arts.

155

NOTES

1. "On Methods of Evolutionary Biology and Anthropology," *American Scientist* 45 (1957): 390. On biology in general, see Peter B. Medawar and Jean S. Medawar, *The Life Science: Current Ideas of Biology* (New York: Harper, 1977).

2. Michael J. S. Dewar, "Quantum Organic Chemistry," *Science* 187 (March 21, 1975): 1037–44.

3. Michael Ruse, *The Philosophy of Biology* (London: Hutchinson University Library, 1973), ch. 4. The theory of plate tectonics plays a similar role in the earth sciences.

4. E.g., Jacques Monod, *Chance and Necessity: An Essay on the Natural Philosophy of Modern Biology,* trans. Austryn Wainhouse (New York: Knopf, 1971), pp. 144–46.

5. The Crab Nebula was not associated with the 1056 supernova until 1928, by the American astronomer Edwin Hubble.

6. *Personal Knowledge*, p. 137.

7. Abraham Kaplan, *The Conduct of Inquiry* (San Francisco: Chandler, 1964), pp. 50–51: "A natural grouping is one which allows the discovery of many, and more important, resemblances than those originally recognized. . . . It is artificial when we cannot do more with it than we first intended."

8. Carl G. Hempel, *Aspects of Scientific Explanation and Other Essays in the Philosophy of Science*, pp. 146–47.

9. Short for "Special Unitary Group in 3 dimensions." "Spin," briefly, is the angular momentum associated with the rotation of an elementary particle on its axis.

10. "Quark" is a word used by James Joyce in *Finnegan's Wake*: "Three quarks for Muster Mark!" Originally it meant "to croak." For information on quarks, see Sidney D. Drell, "Elementary Particle Physics," *Daedalus* 106 (Summer 1977): 15–32.

11. See Romano M. Harré, *The Principles of Scientific Thinking*, pp. 137–41.

12. The word "law" is used to mean both a regularity and a statement intended to describe it, i.e., a law statement. When we say that things and events "observe" or "obey" the laws of nature, we speak metaphorically without realizing it. Considered as regularities, the laws of nature express rather than constrain the characters of things; considered as statements, they describe rather than prescribe how things behave. This vestigial metaphor is a remnant of the belief that nature obeys the laws of God.

13. Bernhard Rensch, *Biophilosophy*, trans. C. A. M. Sym (New York: Columbia University Press, 1971), pp. 131–42.

14. Harré, *Principles of Scientific Thinking*, p. 152.

156

15. The Hardy-Weinberg law states that under certain idealized conditions the genotype frequencies of the zygotes (progeny) are the square of the gene frequencies of the gametes (mature germ cells).

16. On the conventional status of Newton's laws, see Norwood Russell Hanson, *Patterns of Discovery*, ch. 5; and Ernest Nagel, *The Structure of Science*, pp. 174–202.

17. See Harré, *Principles of Scientific Thinking*, pp. 143–45.

18. On laws, see Mario Bunge, "Scientific Laws and Rules," in *Contemporary Philosophy: A Survey*, ed. Raymond Klibansky: vol. 2, *Philosophy of Science* (Florence, Italy: La Nuova Italia Editrice, 1968), pp. 128–40; and John Oulton Wisdom, *Foundations of Inference in Natural Science* (London: Methuen, 1952), pp. 11–26. For a more complete discussion, see R. B. Braithwaite, *Scientific Explanation* (Cambridge: Cambridge University Press, 1953), ch. 9; Harré, *Principles of Scientific Thinking*, chs. 4, 5; and Peter Achinstein, *Laws and Explanation: An Essay in the Philosophy of Science* (Oxford: Oxford University Press, 1971).

19. Ernest Nagel, *The Structure of Science*, pp. 108–9. The logical empiricist view of theories may be found in R. B. Braithwaite, *Scientific Explanation*, chs. 1–3; Rudolf Carnap, "The Methodological Character of Theoretical Concepts," in *The Foundations of Science and the Concepts of Psychology and Psychoanalysis*, ed. Herbert Feigl and Michael Scriven, Minnesota Studies in the Philosophy of Science, vol. 1 (Minneapolis: University of Minnesota Press, 1956), pp. 38–76; Herbert Feigl, "The Origin and Spirit of Logical Positivism," in *The Legacy of Logical Positivism: Studies in the Philosophy of Science*, ed. Peter Achinstein and Stephen F. Barker (Baltimore: John Hopkins University Press, 1969), pp. 15–18; Carl G. Hempel, *Aspects of Scientific Explanation*, ch. 8.

20. Carl G. Hempel, "The 'Standard Conception' of Scientific Theories," in *Analyses of Theories and Methods of Physics and Psychology*, ed. Michael Radner and Stephen Winokur, Minnesota Studies in the Philosophy of Science, vol. 4, p. 152.

21. Harré, *The Principles of Scientific Thinking*, ch. 2. Note that a mechanism is nothing specifically mechanical. See Harré and P. F. Secord, *The Explanation of Social Behavior* (Oxford: Blackwell, 1972), p. 70: "Chemists discover reactions, and by describing the interplay and interchange of ions they explain them. Geneticists discover non-random patterns in the distribution of animal and plant characteristics from generation to generation, an example of which are the patterns described in Mendel's Laws, and they explain the existence and persistence of these patterns by the mechanisms of gene transfer, dominance, and recessiveness, and so on. Evolution is explained by random mutation and natural selection, diffraction is explained by the interference between waves, the aurora is explained by the mechanisms of ion formation high in the rarefied regions of the atmosphere, by electrons from the sun being drawn off toward the poles by the earth's magnetic field, and so on."

22. Harré, *Principles of Scientific Thinking*, pp. 2–3, 60.

23. There are other views of what a scientific theory is (e.g., those of Toulmin, Hanson, Kuhn, Feyerabend, and Lakatos). However, only one other school of thought (represented by Patrick Suppes, Frederick Suppe, and Bas van Frassen) has sought to analyze the structure of a theory in any detail. See Frederick Suppe, "The Search for Philosophic Understanding of Scientific Theories," in *The Structure of Scientific Theories*, ed. Frederick Suppe.

24. This classification is proposed by Peter Achinstein, *Concepts of Science*, chs. 7, 8.

25. Nagel, *Structure of Science*, pp. 401–6.

26. *Ibid.*, p. 405.

27. Francisco J. Ayala, "Teleological Explanations in Evolutionary Biology," *Philosophy of Science* 37 (March 1970): 12.

28. See Ernst Mayr, "Teleological and Teleonomic, A New Analysis," in *Methodological and Historical Essays in the Natural and Social Sciences*, ed. Robert S. Cohen and Marx W. Wartofsky, Boston Studies in the Philosophy of Science, vol. 14 (Dordrecht, Holland, and Boston: Reidel, 1974), p. 110. On teleology in general see—in addition to Ayala, "Teleological Explanations"—J. V. Canfield, ed., *Purpose in Nature* (Englewood Cliffs, N.J.: Prentice-Hall, 1966); David Hull, *Philosophy of Biological Science* (Englewood Cliffs, N.J.: Prentice-Hall, 1973), ch. 4; Michael Ruse, *Philosophy of Biology*, ch. 9; William C. Wimsatt, "Teleology and the Logical Structure of Function Statements," *Studies in History and Philosophy of Science* 3 (February 1972): 1–80.

29. Ernest Nagel, "Issues in the Logic of Reductive Explanation," in *Mind, Science, and History, Contemporary Philosophic Thought: The International Philosophy Year Conferences at Brockport*, ed. Howard E. Kiefer and Milton K. Munitz (Albany: State University of New York Press, 1970), pp. 117–37.

30. See Karl R. Popper, "Scientific Reduction and the Essential Incompleteness of All Science," in *Studies in the Philosophy of Biology: Reduction and Related Problems*, ed. Francisco José Ayala and Theodosius Dobzhansky (Berkeley and Los Angeles: University of California Press, 1974), p. 283.

31. Attempts at reduction encounter other problems. For example, in order to reduce thermodynamics to statistical mechanics, we must identify key thermodynamical terms with statistical ones. But the statistical terms usually designate a number of alternative concepts based on different statistical and other procedures. In Boltzmann's version of statistical mechanics, for example, the equilibrium values of a gas are associated with "most probable" values of certain functions of microscopic states (i.e., of the gas molecules), whereas in Gibbs's version they are associated with "average" values over all possible macrostates. Since different alternative concepts have different advantages, we do not know which to prefer and so which sense of the statistical term to identify with a thermodynamical term. See Lawrence Sklar, *Space, Time, and Spacetime* (Berkeley and Los Angeles: University of California Press, 1974), pp. 379–94.

32. See Thomas Nickles, "Two Concepts of Intertheoretic Reduction," *Journal of Philosophy* 70 (April 12, 1973): 181–201; Lawrence Sklar, "Types of Inter-Theoretic Reduction," *British Journal for the Philosophy of Science* 18 (August 1967): 109–24.

33. See Francisco José Ayala, "Introduction," in *Studies in the Philosophy of Biology*, ed. Ayala and Dobzhansky, pp. vii–xi.

34. Theodosius Dobzhansky, "Introductory Remarks," *ibid.*, p. 1.

35. See David L. Hull, "Reduction in Genetics—Biology or Philosophy?" *Philosophy of Science* 39 (December 1972): 491–99, and *Philosophy of Biological Science*, ch. 1; Ruse, *Philosophy of Biology*, ch. 10. See also three contributions to *PSA 1974: Proceedings of the 1974 Biennial Meeting, Philosophy of Science Association*, ed. R. S. Cohen, G. A. Hooker, A. C. Michalos, and J. W. Van Evra, Boston Studies in the Philosophy of Science, vol. 32 (Dordrecht, Holland, and Boston: Reidel, 1976): Ruse, "Reduction in Genetics," pp. 633–52, now argues that classical genetics can be reduced to Mendelian genetics, as does Kenneth F. Schaffner, "Reductionism in Biology: Prospects and Problems," pp. 613–32; Hull, however, denies this, "Informal Aspects of Theory Reduction," pp. 653–70.

36. Ayala, "Introduction," p. xi: "It is clear that in the current state of scientific development a majority of biological concepts such as cell, organ, Mendelian population, species, genetic homeostasis, predator, trophic level, etc., cannot be adequately defined in physicochemical terms."

37. Some extreme reductionists in this sense are Francis Crick, *Of Molecules and Men* (Seattle: University of Washington Press, 1966); Jacques Monod, *Chance and Necessity*; and Peter Medawar, "A Geometric Model of Reduction and Emergence," in *Studies in the Philosophy of Biology*, ed. Ayala and Dobzhansky, pp. 57–64. An extreme *anti*reductionist is Michael Polanyi. See his "Life's Irreducible Structure," *Science* 160 (June 21, 1968): 1308–12.

38. See Gerald M. Edelman, "The Problem of Molecular Recognition by a Selective System," in *Studies in the Philosophy of Biology*, ed. Ayala and Dobzhansky, pp. 45–56.

39. Henri Poincaré, quoted by J. Hildebrand, *Science in the Making* (New York: Columbia University Press, 1957), p. 89.

159

Chapter 7

The Scientist as a Person

THE PSYCHOLOGY OF THE SCIENTIST

I now turn from the activities and products of research to the human actors in the drama of science. How dramatic science is, and how deeply the personality of the scientist enters into his work, we shall shortly see. Like other people, scientists are driven by strong emotions; each has a unique personality and life history; each has his own needs and interests. In any piece of research it is this unique person who intuits, reasons, experiments, and draws conclusions. Science is a disciplined enterprise seeking impersonal truth, but it also can be highly personal, even subjective.

If this statement sounds paradoxical, it is only because many people suppose that reason and passion are mutually exclusive. Sometimes they are. But often they support each other, as in creative thinking, where emotion provides the driving force and reason the discipline. In science, as elsewhere, reason is exercised *with* emotion, in countless moods, by countless temperaments, in ever-changing social environments. For impersonal truth is not found impersonally but by an immense effort of the whole person. Only the universally accepted findings of science are stripped of emotion, not the struggle to attain them.

Let me illustrate. The Italian physicist, Enrico Fermi, who was notoriously reticent, suffered deeply from his failure to explain the behavior of the transuranic elements (elements with atomic numbers greater than that of uranium). Only the discovery of nuclear fission led to a solution of the problem. Shortly after World War II he and some colleagues were studying the architect's sketches of the future Institute for Nuclear Science at the University of Chicago. The drawings showed a vaguely outlined human figure in bas-relief over the

160

entrance. When his group began speculating what the figure represented, Fermi suddenly blurted out that it probably was "a scientist *not* discovering fission!" [1]

Scientists are extraordinarily varied; they differ in thought, manners, morals, temperament, and purpose in life. In any group of scientists there are extroverts and introverts, altruists and self-seekers, dreamers and realists, believers and atheists. Does this mean that psychologists cannot begin to classify them? Not at all. In fact, the basic divisions of scientific research tend to correspond to certain broad personality types. [2]

PERSONALITY TYPES

Scientists have been classified by personality type into theorizers, empiricists, and intermediates. [3] Theorizers generally are adventurous thinkers with an urge to challenge accepted ideas. (But not always—Max Planck, the originator of quantum thinking, was a most reluctant revolutionary.) Often they are highly inventive, producing one hypothesis after another. Einstein ranged across physics, opening new paths in mechanics, electromagnetism, quantum theory, gravitation, and the unified field. Clerk Maxwell wrote a succession of brilliant papers in both kinetic theory and electromagnetism. Theorizers tend to be deeply committed to their ideas, and will often defend them aggressively. To reach a wider public, Galileo wrote in Italian rather than Latin and made devastating fun of his scientific and ecclesiastical opponents. Yet some theorizers have let their ideas speak for themselves. At Yale, Willard Gibbs, who was a courteous, self-contained man, did not attend scientific meetings and did almost nothing to advertise his highly original work in thermodynamics and statistical mechanics. As a result he remained virtually unknown in this country. But whether they promote their ideas or not, theorizers usually have great confidence in them, a confidence that helps them ride out the opposition that original thinking so often encounters. Indeed, some theorizers have even "improved" the evidence to support hypotheses of which they were convinced. Both Dalton and Mendel published data that were too good to be true, lacking the statistical fluctuations that must have occurred. Ptolemy seems to have been a fraud on a grand scale, inventing and doctoring data throughout his theory.[4]

Empiricists usually are methodical and ultra-careful. Their energies are absorbed in observing, experimenting, and measuring phenomena with utmost precision. They tend to be strongly committed to existing theories, and generally are reluctant to invent hypotheses. Albert Michelson, for instance, was dismayed by the failure of his first experiment (1881) to detect any sign of an ether. Four years later, as he was beginning his second experiment, he had a nervous breakdown so severe that his co-worker, Edward Morley, thought that he never would do research again. Michelson disliked the theory of relativity to the end. In 1931 Einstein, hearing that Michelson was seriously ill, went to call on him, and was warned by his daughter not to mention the ether.[5] Some empiricists, however, have been bolder. Carl Anderson declared that he had found evidence of a positively charged electron when, almost to a man, physicists believed this to be impossible.

For empiricism in its purest form, however, we turn to the nineteenth-century German experimentalist, Friedrich Kohlrausch. This man lived to measure. He had a mania for counting steps and thinking of ways to save time, so that he could return to his desk and pore over his numbers. He began his career in the 1860s just as German laboratory physics was about to expand. His course in precision measurement at Göttingen was widely imitated. But he persuaded best by example. He told his students that no physical fact was too low or trifling to measure. He would be delighted, he said, to measure the speed of water in a gutter. He was well rewarded for his dedication, becoming Hermann Helmholtz's successor as president of the Physical-Technical Institute, the most prestigious position in German physics.[6]

Finally, there are scientists who are competent at both forms of work. Intermediates can think boldly, but as a rule they prefer to work within an established tradition. Rutherford and Fermi are outstanding examples. Rutherford was primarily an experimentalist, Fermi a theorist. Both thought boldly but not speculatively. A rather

162

different example is the Austrian-born physicist Paul Ehrenfest, who was a critic rather than an experimentalist. He would point out the least mistake swiftly but goodnaturedly. Einstein and Bohr valued his criticism so highly that they often discussed their work with him. But his critical faculty acted as a brake on his imagination, and he suffered deeply from his failure to do theoretical work as good as that which he criticized. In 1933 he committed suicide.[7]

The Discovery of the Structure of DNA These contrasting, psychologically types are strikingly exemplified by the scientists most closely involved in the discovery of the structure of the DNA molecule.[8] Two main teams competed in the race to make this discovery: Francis Crick and James Watson at the Cavendish Laboratory, Cambridge, and Maurice Wilkins and Rosalind Franklin at King's College, London. Crick and Watson were great speculative scientists. Talkative, quick-thinking, self-assured, Crick teemed with new ideas, which he liked to develop in discussion. Like many brilliant men, Crick was apt to upset people by pointing out the implications of their ideas before they did. He annoyed his director, Sir Lawrence Bragg, by sharply querying some of his ideas, and as a result he and Watson had to pursue their project despite Bragg's disapproval. If Crick gave the partnership vitality, Watson brought it staying power. When Crick returned temporarily to his dissertation, Watson remained with DNA and continued to pursue the crucial idea that the molecule was made of two chains of chemical bases rather than three. Less ebullient than Crick, Watson was sustained by his passionate resolve to make the discovery before anyone else.

Between the Cambridge men and their competitors in London, what a world of difference! Rosalind Franklin epitomized the empirical scientist. A superb experimenter, she analyzed the X-ray photographs that yielded much of the evidence needed to unravel the structure of DNA. But she refused to leap beyond the evidence and so missed making the crucial discovery. As soon as Watson saw one of her photographs, he realized that it indicated a helical structure, and he worked out the two-chain hypothesis at once. Franklin, however, disdained such activity as an attempt to prejudge the hard work of experimentation that in time would make the underlying structure unequivocally clear.

Franklin's partner, Maurice Wilkins, was a scientist of the intermediate kind. Less confident than Crick or Watson, he was unwilling to take risks. Although he had worked on DNA for several years (in England it was regarded as "his" problem), he had yet to propose a comprehensive hypothesis. He also was unable to cope with Franklin's aggressive, argumentative manner. (Franklin thought that her work was undervalued because she was a woman, and she reacted by asserting herself the more strongly.)[9] Instead of talking problems through with her, Wilkins withdrew. The feud sapped his energy. A more vigorous and self-assured individual would have mastered both the personal and the scientific problem.

Mention also should be made of Fred Griffith, another empiricist who made a vital contribution to the Watson-Crick discovery. Griffith is ignored in Watson's book, but as Peter Medawar has pointed out, "No Fred, no Jim."[10] Griffith discovered "transformation," a phenomenon that occurs when a chemical (now known to be DNA) from one type of microbe causes a permanent change in the behavior of another type. Griffith announced his discovery in 1928, when it was commonly believed that proteins carried the coded instructions in living cells, DNA being thought too small a molecule for this role. (It was not until 1944 that Oswald Avery, MacLeod, and McCarty confirmed that DNA was indeed the chemical responsible.) Quiet, and retiring, Griffith avoided scientific societies. Avery, the leader in microbiology, neither knew him nor corresponded with him. Griffith worked steadily and meticulously by himself, doing beautifully designed experiments, and grudging even the time spent writing papers. An infinitely cautious man, he reached the hypothesis of transformation against his own deep-seated convictions, and in his paper refused to speculate whether DNA had transformed the bacteria. His reputation as the most sober of experimentalists ensured that his paper was noticed.

The DNA discovery also illustrates the part played in research by personal relations. Wilkins was delayed by his quarrel with Franklin. Watson, on the other hand, was stimulated to press on with his helical hypothesis when Wilkins generously showed him Franklin's photographs—something Wilkins need not have done. Watson, again, was saved from pursuing an incorrect version of his hypothesis by Jerry Donohue, a structural chemist, who told him that identical bases in corresponding chains could not be bonded by hydrogen. Donohue had worked for many years under Linus Pauling, who also was seeking the structure of DNA at Caltech in Pasadena. Ironically, Donohue's advice probably enabled Watson and Crick to beat Pauling to the discovery. So much for the myth that great science is detached and impersonal!

PSYCHOLOGICAL FACTORS IN RESEARCH

Certain specific psychological factors are also expressed in a scientist's work, among them his personality traits, his deep convictions, style of thought, subconscious thinking, unconscious motives, and power of perception.

Traits and Convictions: Einstein and Planck. Einstein combined a remarkable number of contrary traits. He loved mankind in general but was close to no one, not even his wife. He once admitted: "My passionate interest in social justice and social responsibility has always stood in curious contrast to a marked lack of desire for direct association with men and women. I am a horse for a single harness, not cut out for tandem or teamwork."[11] Einstein always had been solitary. At school he remained aloof, absorbed in books or music. Later he wrote: "I live in that solitude which is painful in youth but delicious in the years of maturity."[12] Again, though he believed in democracy, he held that most men were incapable of thinking for themselves. He shunned politics and publicity yet urged Roosevelt to build an atom bomb. He wrote papers that are extraordinarily clear and logical, but he maintained that basic principles must be intuited. 'To these elementary laws," he wrote, "there leads no logical path, but only intuition, supported by being sympathetically in touch with experience."[13] Einstein was well aware of such polarities in himself.

Einstein's capacity to contain polarities was matched by, and

165

expressed in, his ability to explore and resolve contradictions in scientific thought.[14] Each of his three great papers of 1905—on the photoelectric effect, Brownian motion, and special relativity—addresses a conflict between theories rather than between theory and fact. The first paper treats the conflict between Planck's quantum hypothesis and Maxwell's electromagnetic theory; the second, that between thermodynamics and statistical mechanics; the third, that between Maxwell's theory and classical mechanics. Unlike many of his contemporaries (Hendrik Lorentz and Max Abraham, for example), Einstein sought to reconcile mechanics with electrodynamics rather than reduce the one to the other. In his theory of general relativity he combined mechanics with gravitation theory, and in his unified field theory he endeavored to unite gravitation with electrodynamics.

In seeking to unite theories, Einstein also was following his deep-rooted belief in the ultimate harmony of the universe and in the capacity of our theories to reflect this harmony with a simplicity of their own. This belief he called the "cosmic religious feeling, the strongest and noblest motive for scientific research."[15] Though contemptuous of the established faiths, he was, as he said, "a deeply religious unbeliever."[16] This attitude seems to have sprung from an incident in early childhood, his awed observation that a compass needle always swung round to point in the same direction. Much later he wrote, "That this needle behaved in such a determined way did not at all fit into the nature of events, which could find a place in the unconscious world of concepts. . . . I can still remember—or at least I believe I can remember—that this experience made a deep and lasting impression on me."[17]

Thus in 1907 Einstein was not disturbed to learn that two other theories (those of Max Abraham and A. H. Bucherer) agreed more closely than the theory of special relativity with certain experimental findings. He maintained that because his theory was more simple and comprehensive, it was more likely to represent the order of nature, and that a decision between the theories should be delayed until more facts were known. Ten years later the data were found to be mistaken.

As another instance, take the confirmation of general relativity. In 1919, to test the prediction that light rays are deflected by the sun's gravitational field, Arthur Eddington photographed a total eclipse off the coast of West Africa. Einstein did not rush to open Eddington's telegram. When asked what he would have done if the test had gone against him, he replied, "Then I would have been sorry for the dear Lord—the theory is correct."[18] Nevertheless, this man of contraries was also deeply moved, for he wrote to Planck, "It is a gift from fate that I have been allowed to experience this [event],"[19] and, to Eddington, that "if it were proved that this effect does not exist in nature, then the whole theory would have to be abandoned."[20]

Yet again, from 1926, when Heisenberg formulated the uncertainty principle (that one cannot at the same time measure exactly both the position and the momentum of an elementary particle) until his death nearly 30 years later, Einstein vigorously resisted the non-determinist Copenhagen interpretation of quantum mechanics. The fundamental laws, of nature, he declared, cannot be laws of chance. To Max Born he wrote:

> Quantum mechanics demand serious attention. But an inner voice tells me that this is not the true Jacob. The theory accomplishes a lot, but it does not bring us close to the secrets of the Old One. In any case, I am convinced that He does not play dice.[21]

By "God" or "the Old One" he meant the order of nature. To Niels Bohr he sent a succession of apparent paradoxes that he had found in the theory. Whenever Bohr solved one, Einstein would think of another. "God," he maintained, "is sophisticated, but not malicious" ("Raffiniert ist der Herr Gott, aber boshaft ist Er nicht"). (I would translate "raffiniert" as "sly" or "subtle.")

Einstein was a natural nonconformist. He wore his hair long and refused to have his suits pressed. As a student he was undistinguished. Unable to secure a permanent post as a schoolteacher, he lived precariously from temporary teaching jobs. Through the efforts of a friend's father he was given an assessor's position at the Swiss Patent Office in Berne. When appointed to the University of Prague, he held discussions with his first pupil, Otto Stern, also a future Nobel Laureate, in a café attached to a brothel.

What a contrast with Planck! Unlike the freethinking, wandering Einstein, Max Planck was an ardent patriot and a deep conservative. Born into an old and distinguished German family, he strove to uphold the morality and manners of his class. In his dark suit, starched white shirt, and black bow tie he looked like a high-ranking civil servant. He lectured precisely and dryly, and was slow to commit himself to an opinion. One of his students, the future atomic physicist Lise Meitner, said that to all her questions he replied, "I will give you my answer tomorrow." In his long life (he died at 89) he bore much misfortune with great dignity. He lost his first wife in 1909 and a son in each of the World Wars. Two daughters died in childbirth. In 1944 his manuscripts and most of his books were destroyed in an air raid.

As a child Planck had revealed a certain talent for music, but he may have given up the idea of a musical career when told by a professional musician, whom he had consulted, "If you have to *ask*, then you'd better study something else!" At any rate, between Einstein and Planck—otherwise so different—a strange friendship grew, based on physics, about which they usually disagreed, and music, on which they thought alike. They would spend hours together playing chamber music, Einstein on the violin and Planck at the piano.

A descendant of lawyers who had helped shape the Prussian legal code, Planck revered the laws of God and man alike. He believed that the laws of nature reveal the perfection of God. He was deeply committed to classical physics. He was guided also by the conviction that science should be the search for the absolute, by which he meant the objective laws and physical constants (e.g., the velocity of light, the charge and rest mass of an electron) of a universe independent of man. As he himself put it,

> My original decision to devote myself to science was a direct result of the discovery which has never ceased to fill me with enthusiasm since my very early youth—the comprehension of the far from obvious fact that the laws of human reasoning coincide with the laws governing the sequences of impressions we receive from the world about us; that, therefore, pure reasoning can enable man to gain an insight into the mechanism of the latter. In this connection, it is of paramount importance that the outside world is something independent from man,

something absolute, and the quest for the laws which apply to this absolute appeared to me as the most sublime scientific pursuit in life.[22]

From 1920 till his death he was a churchwarden, professing to believe in an almighty, omniscient, beneficent God, though not a personal one. He also wrote on religion and philosophy. Science strives toward God, he declared, because it seeks the absolutes that He has created.

Planck's absolutism and conservatism are exemplified in his struggle with the black body problem and the quantum hypothesis. Early in his career he worked on the application of the laws of thermodynamics, holding these laws to be "absolute" because they were universal in scope and unaffected by particular circumstances. (Even as an old man he recalled his *Gymnasium* teacher's account of the law of the conservation of energy.[23]) He then chose to investigate the problem of black-body radiation—radiation of all frequencies emitted by a so-called black body or perfect reflector which absorbs all radiation reaching it. This radiation is independent of the properties of any material bodies that are in equilibrium with it. For Planck, then, it represented "something absolute." The problem was to provide a theoretical explanation of Wilhelm Wien's law linking the frequency, intensity, and temperature of such radiation. Planck first tried and failed to derive the law from Maxwell's electromagnetic theory. Next he deduced it from the laws of thermodynamics but was dissatisfied with the result. Then, seeing no alternative, he turned to Boltzmann's statistical mechanics, a theory he found deeply distasteful. (This theory formulates the thermodynamic laws statistically, and hence less than absolutely, so that they hold only more or less probably in the individual case.) Yet even then, during "a few weeks of the most strenuous work of my life," he met another setback.[24] To his horror the energy of the system he was studying turned out to consist of a finite number of discrete elements or "quanta," a clear contradiction of the Maxwellian demand for a continuous, infinitely divisible quantity. Finally, in what he called "an act of desperation," resorted to because "a theoretical explanation [of Wien's law] had to be supplied at all costs,"[25] he proposed the revolutionary hypothesis

169

that the energy of electromagnetic radiation exists in discrete quanta. But at least he had discovered a universal constant, the quantum of action.

Striving to uphold classical physics, he spent the next 15 years trying to reconcile his hypothesis with Maxwell's theory. As he declared:

> My futile attempts to fit the elementary quantum of action somehow into the classical theory continued for a number of years, and they cost me a great deal of effort. Many of my colleagues saw in this something bordering on a tragedy. But I feel differently about it. For the thorough enlightenment I thus received was all the more valuable. I now knew for a fact that the elementary quantum of action played a far more significant part in physics than I had originally been inclined to suspect.[26]

Planck also ignored Einstein's extension of the quantum hypothesis in his explanations of the photoelectric effect (light, said Einstein, travels in quanta), believing it too soon to abandon Maxwell's theory. However, he welcomed the theory of relativity. In denying the absolute nature of space and time, the theory does not, he argued, eliminate the absolute but locates it in the metric of the space-time continuum, a metric that is independent of the measuring processes and reference frames that determine space and time.[27]

Styles of Thought: Faraday and Maxwell. Every scientist also has his own style of thought. Some scientists, such as Paul Dirac, seek the simplest mathematical theory. Others start from philosophic ideas, as Heisenberg later worked from ideas in Plato. Others still, such as Rutherford, need a physical picture. Let us contrast Michael Faraday with James Clerk Maxwell.

Faraday thought pictorially and intuitively. He formed a number of vague but powerful metaphysical ideas, such as the idea that all matter is ultimately force. His belief in these ideas, which were ignored by the scientific establishment, was strengthened by his faith in God as creator and sustainer of the universe. (Faraday be-

170

longed to the Sandemanians, a Protestant sect believing in love and community.) He considered himself an instrument through which truth would be revealed. With the aid of mental images he invented a range of high-level theories to exemplify these ideas and then imagined and carried out experiments to test them.[28] A natural experimenter, he had become skilled with his hands during his seven years as a bookmaker's apprentice. "A good experiment," wrote physicist John Tyndall, "would make him almost dance with delight."[29] He worked tirelessly, recording his progress daily in his carefully cross-referenced diary. His motto? "Work, finish, publish." Sometimes he passed from experiments to theories by picturing the invisible entities and interactions from which the phenomena he had observed might ensue. His experiments convinced him that electric and magnetic forces could not act instantaneously at a distance but must travel from point to point over time through a continuous medium.

In a series of experiments beginning in 1831, Faraday showed that a moving magnet produces an electric current. Thus he complemented Hans Christian Oersted's discovery, a decade earlier, of the reverse effect. In his simplest experiment Faraday pushed an iron bar magnet through a wire coil attached to a galvanometer and found that an electric current ran into the coil. To explain this result, he reasoned pictorially from another observed phenomenon—that iron filings, strewn over a card placed on a magnet, arrange themselves in lines. From this fact he generalized that "lines of force" exist in the space surrounding a magnet whether iron filings are placed there or not. These lines form a field of influence whose strength can be measured at any point by observing the behavior there of a compass needle or iron filing. He pictured the lines as invisible, elastic threads stretched by tension and separated by pressure. According as they converge or spread apart, the field is strong or weak. An electric current is created, he proposed, whenever a circuit is cut by lines of magnetic force, as when a magnet is moved through a wire coil.

What was needed now was a precise account of how the fields of force behave. Faraday, however, was self-taught (he began his apprenticeship at 14 after only slight schooling), and he was not well versed in mathematics. Hence he was unable to formulate a set of exact laws describing the interactions of the forces he had portrayed.

171

These laws were provided by Maxwell, who combined mathematical ability with unusual insight into nature.

Maxwell was a Scottish laird (lord), thoughtful, reserved, and deeply conscious of his responsibility to others. He nursed his wife devotedly, even when ill himself. During one period of three weeks' nursing he did not sleep in a bed, though he continued to lecture and work as usual. He, too, was deeply religious but kept his religious thoughts to himself. Invited to join a society of scientists for the discussion of religious questions, he replied, "I think that the results which each man arrives at in his attempts to harmonize his science with his Christianity ought not to be regarded as having any significance except to the man himself and to him only for a time." He also had a gentle sense of humor. On a visit to Cambridge from Aberdeen, where he was professor, he brought with him a favorite spinning top. One evening he showed it to friends, who left it spinning in his room. Next morning he saw one of them coming across the court. So he jumped out of bed, started the top, and retired again, pretending to be asleep. Apparently his friend was taken in.

Maxwell studied mathematics at Edinburgh Academy under James Gloag, a man of "strenuous character and quaint orginality" to whom "mathematics was a mental and moral discipline."[30] Throughout his career he preferred geometry to algebra because it focused his attention on conceivable physical entities. At Cambridge he was a student of the great teacher of mathematical analysis, William Hopkins. But even in Hopkins's lectures, as a fellow student remarked,[31] "whenever the subject admitted of it, he [Maxwell] had recourse to diagrams, though the rest [of the class] might solve the question more easily by a train of analysis." Hopkins, according to a witness, called him "unquestionably the most extraordinary man he has met with in the whole range of his experience; he says it is impossible for Maxwell to think incorrectly on physical subjects; that in his [study of] analysis, however, he is far more deficient."[32]

Maxwell, then, refused to theorize by mathematics alone (in which, he said, "we entirely lose sight of the phenomena to be explained") but used mathematics together with a clear physical idea of the entities he postulated. Yet, as James Jeans points out, "he was perhaps less remarkable in the possession of a vivid physical imagination than in the strict control he kept over it. He never allowed it to run away with him."[33] This was partly due to his careful use of analogies. In Maxwell's words, "We must . . . discover some method of investigation which allows the mind at every step to lay hold of a clear physical conception, without being committed to any theory founded on the physical science from which that conception is borrowed."[34] An analogy is a partial resemblance. It can help us to think creatively, provided we do not regard the things analogized as being identical. As Maxwell observed, it is helpful to consider an electric current as analogous to phenomena as different as heat con duction and fluid motion, "so long as we do not let the analogies suggest to us that electricity is either a substance like water, or a state of agitation like heat."[35] The analogy in this case is geometrical, "a similarity between [mathematical] relations, not between the things related."[36] Thanks to his method of analogies, Maxwell had many ideas but used them carefully.

How did Maxwell solve Faraday's problem? First he worked out an analogy between Faraday's lines of force and streamlines in a fluid. Then in a second paper he devised an elaborate mechanical model of an electromagnetic field, from which he derived a set of partial differential equations describing the field's behavior. He represented Faraday's line of magnetic force as a string of spinning cells or vortices and his electric current as a layer of tiny particles, like ball bearings, surrounding each vortex. When a vortex changes its speed of rotation, it causes some or all of the surrounding particles to move to the next vortex, thus producing an electric current. The particles cause this vortex in turn to change its speed of rotation, thus extending the current.

Here was a mechanical analogy to the process by which a changing magnetic field creates an electric current (or field) and a changing electric current (or field) creates a magnetic field. Reflecting on the action of the vortices, Maxwell derived an equation that expressed quantitatively the relation, proposed by Faraday, between the tension along the lines of force and the pressure between them. He then related the motion of the particles to the rotation of the vortices, and proposed equations relating the variation of a magnetic field to variations in the electric current or field it generates. By

173

means of these derivations and others, Maxwell put together a set of equations correlating electric and magnetic variations, such as the variation of an electric field in space with the variation in time of a magnetic field. These equations state in precise terms the laws of electromagnetic behavior which Faraday had been seeking. In a third paper Maxwell dropped the model but retained the equations, stating only that the equations described the behavior of some currently unknown mechanism obeying Newton's laws.

Nonconscious Thought Processes. I now turn to some more dimly lit areas of the scientist's mind. Nonconscious processes are known to play an important part in scientific work, especially in the invention of hypotheses, though just what that part is remains obscure.[37] Not a few scientists have said that a major hypothesis—a solution to a problem they have meditated on for some time—has come to them suddenly in a moment of illumination, after a long period of conscious preparation followed by an interlude of unrelated activity in which they have not consciously thought about the problem at all. First, it seems, the scientist has an insight into his problem, which he follows up, inventing and investigating one or more solutions, none of which satisfies him. Getting nowhere, he sets the problem aside but continues to work on it subconsciously, freed from the constraints of literalistic thinking. Eventually, if he is lucky, this work produces a solution, which then "occurs" to him.[38]

Consider a well-known case. August Kekulé realized that the carbon atoms in organic molecules are linked in chains but he was unable to explain the structure of the benzene molecule, which contained six carbon atoms. The solution came to him in a waking dream. As he sat by the fire, carbon atoms began to dance before his eyes and form chains like snakes. Suddenly one of the snakes seized hold of its tail to form a ring. At once Kekulé realized that in the benzene molecule the chain of carbon atoms is a closed one. Thus the fundamental concept of organic chemistry, the benzene ring, was born. Such inspired solutions are not self-certifying, of course, but must be tested like any other hypothesis. As Henri Poincaré observed:

174

It never happens that unconscious work provides *ready-made* the result of a lengthy calculation in which we have only to apply fixed rules. . . . All that we can hope from these inspirations, which are the fruits of unconscious work, is to obtain points of departure for such calculations. As for the calculations themselves, they must be made in the second period of conscious work which follows the inspiration, and in which the results of the inspiration are verified and the consequences deduced.[39]

Insight comes only to those who seek it. Isolated original ideas may be had fairly easily, but an original system of ideas requires a long process of intellectual labor. Howard Gruber has made this point well in his account of the movement of Darwin's thought up to his reading of Malthus and his first formulation of the final theory of evolution.[40] Darwin wrote down his first, or "monad," theory of evolution in July 1837, 15 months before reading Malthus. According to this theory, organisms constantly adapt to their environment. Simple living organisms, or monads, are continually springing into existence from inanimate matter. As they change, they become more complex. Darwin illustrated the monad, with its tendency to vary adaptively, with a diagram of a tree branching irregularly (the branches representing adaptive variations).

This idea was the germ of Darwin's next theory. For the irregular branches of the tree accounted better than the monad principle for the observed discontinuities between species—"the great gap," as Darwin put it, "between birds and mammalia, still greater between vertebrates and articulata, still greater between animals and plants."[41] So he abandoned the monad theory and proposed that species are continually giving birth to new species in an unbroken succession. Thus the branching tree of life is a tree of perpetual becoming.

Darwin began to realize, however, that variations are not always adaptive. If they were, there would be no need for natural selection. So he gave up his second theory and treated the fact of variation as an unexplained premise, not a consequence of environmental change. As Gruber puts it, "Evolution through natural selection *means* that variations are not necessarily adaptive, so that varia-

175

tion must be coupled with selection if adaptive change is to occur." At the same time that he realized the extent of variation in nature, he also was struck by nature's fertility. Each organism, he observed, tends to reproduce more of its kind than the number alive in previous generations. This multiplication of species and organisms leads to the struggle for existence and hence to natural selection. Nevertheless, Darwin still saw natural selection as a conservative force, as one that extinguished species rather than created them. What he needed was the insight that if there are many variations and too many individuals for the available resources, then those individuals carrying favorable variations will survive, and in this way new species will come into being. This was the insight Darwin had on reading Malthus. But unless he had been ready for the insight, Malthus would have meant nothing to him as Malthus meant nothing to Alfred Russel Wallace, who read him 14 years before he proposed his theory. What, then, did Malthus give to Darwin? A clear statement of the law of population growth at the moment when he was prepared to recognize its significance for natural selection.

Unconscious Motives. Like everyone else, the scientist is influenced by unconscious motives. Even rational choices often are partly rationalizations of repressed desires not fully explained until their connection with the unconscious has been made clear. With the exception of Freud's early essay on da Vinci—in which Leonardo's observation of nature is explained as the sublimation of an unconscious desire to gaze on his mother's face—there have been no psychoanalytic studies of the lives of scientists comparable with those on writers and artists. This is unfortunate, because scientists are no less prone to neurosis than men of letters. Boltzmann's suicide, Faraday's breakdowns, and Darwin's illness, all of which either resulted from or influenced their scientific work, call out for psychoanalytic explanation.[42] Perhaps the only major study that attempts to consider the unconscious sources of a scientist's work is Frank Manuel's fascinating *Portrait of Isaac Newton*.[43]

Manuel portrays Newton as an extraordinarily neurotic, tormented, and unpleasant man. He was deserted at the age of 3 by his mother, who went to live with another man and only returned when Isaac was 11. As a result he suffered throughout life from an overwhelming sense of insecurity, which he sought to assuage by seeking power—first over nature through thought, then over his rivals through invective and intrigue, and finally through exercising the full powers of the law (notably the death penalty) over the counterfeiters who came before his judgment as Warden of the Mint. Newton sublimated a desire to know his father, who died before he was born, by investigating nature as the handiwork of the Eternal Father. He sublimated anger at his stepfather for usurping his place with his mother by the immense energy with which he tried to explain nature and dominate it intellectually. He sought security by formulating nature's regularities in mathematical laws. His greatest achievement, the derivation of two laws corresponding to Kepler's first and second laws of planetary motion, was an act of psychic restitution made 6 months after his mother's death. Then in 18 months of intense labor he completed his theory for publication, partly out of affection for Edmund Halley who was encouraging him, and partly to eclipse his hated rival, the brilliant but equally unpleasant Robert Hooke. Newton's notorious reluctance to publish sprang not only from his unwillingness to give of himself to others but also from fantasies of omniscience, which he feared to expose to public criticism. Though his narcissism deterred him from entering into conflict with others, his need for security drove him to defend his work with a ferocity unparalleled in science against a succession of critics whom he hounded and crushed, notably Hooke (whom he succeeded as head of the Royal Society), John Flamsteed (the Astronomer Royal), and even Leibniz. How strange the contrast between the incomparable intellect and the imperfect man. And how rare the gentleness revealed in this confession made late in life:

> I do not know what I may appear to the world; but to myself I seem to have been only like a boy, playing on the seashore, and diverting myself, in now and then finding a smoother pebble than ordinary, whilst the great ocean of truth lay all undiscovered before me.[44]

Perception. How a scientist perceives the phenomena he is studying depends on his senses, his training, his research tradition, and, notably, his choice of hypotheses. A scientist familiar with the insights and techniques of other research traditions is less likely to be rigid in his perceptions than a scientist who operates within the concepts and paradigm cases of a single tradition. Sometimes, of course, scientists misperceive. A wrong identification of a virus or a misreading of an X-ray photograph may have serious consequences, such as the apparent refutation of a promising hypothesis. Sometimes scientists disagree about what they perceive. Sometimes they perceive nothing significant at all. For example, the existence of the positron was deduced theoretically by Paul Dirac in 1928, but the experimental evidence for it was not recognized until 5 years later by Carl D. Anderson.[45] Indeed, Anderson's achievement was all the greater because he interpreted the evidence for himself without having heard of Dirac's prediction. During this time the tracks left by positrons on photographic plates in cloud chambers were dismissed as so much "waste." Scientists resisted the notion of the positron, because they thought that the two forms of electricity, positive and negative, required one particle apiece, and that a third particle therefore would have to belong to some further, quite inconceivable, kind of electricity. Hence, they did not observe positrons because they did not expect to.[46]

SUBJECTIVITY IN SCIENCE

From what I have said so far it should be clear that scientists are driven by powerful emotions. They are strongly committed to their own ideas and may feel equally threatened by opposing ones. We can see this in the reactions of physicists in 1926 to Schrödinger's two papers on wave mechanics, treating quantum phenomena as continuous quantities with the differential equations of classical physics. Berlin physicists, upholders of the classical tradition, rejoiced. They had been depressed by Heisenberg's matrix mechanics, published 6 months earlier, which treated these phenomena as discontinuous and hence openly rejected classical physics. The conservative Planck hailed Schrödinger's theory as a vindication of the classical approach. He is reported to have said, "I am reading it as a child reads a puzzle." Einstein, too, was delighted since he long had sought to unify physics. Others were less struck. Bohr, who had proposed the first quantum theory, based on the notion of discontinuity, argued heatedly with Schrödinger until eventually the latter fell ill. Schrödinger, wrote Heisenberg, declared that "if all this damned quantum jumping were really here to stay, I should be sorry I ever got involved with quantum theory." The usually genial Bohr then became a "fanatic." In Heisenberg's words:

> Although Bohr was normally most considerate and friendly in his dealings with people, he now struck me as an almost remorseless fanatic, one who was not prepared to make the least concession or grant that he could ever be mistaken. It is hardly possible to convey just how passionate the discussions were, just how deeply rooted the convictions of each, a fact that marked their every utterance.[47]

Schrödinger and Heisenberg detested each other's theories. Heisenberg wrote to Wolfgang Pauli that "the more I ponder about the physical part of Schrödinger's theory, the more disgusting [abscheulich] it appears to me." Schrödinger published his opinion of Heisenberg's theory: "I was discouraged [abgeschreckt] if not repelled [abgestossen]."[48]

Nevertheless, this very emotional commitment enables the scientist to develop his ideas fully, often against opposition, and to scrutinize vigorously those ideas he opposes. If scientists were not committed to their own ideas, many promising hypotheses would die on the vine; and if scientists were not committed to different ideas, science would sink into dogma and decline. As I have said, science advances in part through the conflict of research traditions, and so through a process in which scientists holding different ideas develop the implications of their own assumptions, and criticize those of their opponents, as energetically as they can.

Where, then, does the objectivity of science reside? It resides partly in this very mutual criticism, which ensures that theories are scrutinized rigorously. It also lies in the methods the scientist uses to develop and test his own hypotheses, making them internally consistent and weighing them against the evidence, in order to anticipate the searching criticism of his fellow scientists by whom they are to be assessed. Thus the evaluation of a hypothesis by the scientific community is complemented by the preliminary evaluation of the hypothesis by its creator. Because the scientist wants passionately to persuade other scientists of the truth of his hypotheses, he seeks to make those hypotheses as logically sound and as adequate to the facts as he can.[49]

THE SCIENTIST AND THE WIDER WORLD

The scientist also belongs to the world of his time. Here I anticipate my argument in later chapters, for I wish to show how the forces that influence science are felt by the scientist himself. The scientist may be placed in a series of settings: the intellectual setting of the research tradition, the specialty, and the discipline, which provide the knowledge, aims, and techniques from which his research problem arises; the social setting of the scientific community, with its mores, hierarchies, and institutions, within which he is linked by personal relations to a variety of colleagues; the broader setting of the intellectual milieu as a whole with its great wealth of schools and achievements; and the widest setting of all, the sociocultural environment with its countless interconnections, social, political, legal, and economic. Each scientist, then, is the focal point of a vast web of influences which it rarely is possible to disentangle completely.

Helmholtz and Energy Conservation. To appreciate this point, let us see why Hermann von Helmholtz turned out to be the one who formulated the law of the conservation of energy in 1847. Here the right man met the right problem at the right time in the right place. The problem was to reconcile the idea of the conservation of mechanical energy (still confused with "force")—which stemmed from Leibniz and had been used to explain phenomena in electricity, magnetism, and chemistry—with Newtonian mechanics, in which energy was not conserved. During the 1840s a number of scientists were working on this problem, but as Yehuda Elkana has argued, Helmholtz, at age 26, had the best combination of knowledge and talent for solving it.[50]

Helmholtz was an unusually gifted person. An accomplished poet and musician, he had studied mathematics, physics, and philosophy, had been trained as a physician, and now was working in the laboratory of Johannes Müller in Berlin, where the attempt was being made to reduce physiology to physics. Steeped in the philosophical tradition of the German universities and especially in the philosophy of Kant (who had maintained that all natural phenomena can be reduced to the forces of attraction and repulsion), Helmholtz thought that all matter is ultimately force and that force in this general sense is "conserved." Again, as a researcher in reductionist physiology (whose fundamental concept was "force of life") he not only was familiar with apparently nonmechanical forces but believed that they could be explained by physics, which he thought reducible to mechanics (whose fundamental concept was "force"). Finally, from his study of mathematics he was proficient both in the vectorial mechanics of the English Newtonian tradition, with its emphasis on force, and in the analytical mechanics of the Continental tradition (Leibniz, Euler, Lagrange), which stressed mechanical energy and stated that it was conserved.

From this background Helmholtz argued in essence that (1) all sciences are reducible to mechanics and all forces to Newtonian forces; (2) since (as he showed) Newtonian and Lagrangean mechanics are equivalent mathematically, the action of any force can be measured (by Lagrangean methods) in terms of the work it does and the energy it transmits; and (3) all such energy is conserved. From Lagrange's mechanics Helmholtz then derived a general equation for the conservation of energy in the universe. This sweeping but exact generalization, which unified a variety of otherwise mysterious forces (heat, light, magnetism, electricity, gravitation, chemical affinity, vital force) and founded the science of thermodynamics, required from Helmholtz not only great courage and imagination but also a commensurate range of skill and experience.

To realize what Helmholtz gained from his knowledge of

181

mathematics, compare his achievement with that of Faraday. Faraday had as much courage and imagination as any scientist in history. He also had experimented extensively in electricity, magnetism, and chemistry, and he believed firmly in the conservation of force. But although in due course he studied the mathematics of Newtonian mechanics, he knew little about the mathematics of Lagrange. In his essay "On the Conservation of Force," written after he had read Helmholtz's paper in translation, Faraday treated force as the "exertion of power" in some direction. Granted, Helmholtz's words are ambiguous; nevertheless his Lagrangean mathematics make it clear that what is conserved is not force in Faraday's sense but the energy that forces expend. If Faraday had been familiar with Lagrangean mathematics, he would have seen that force as such is not conserved, and he might have realized, like Helmholtz, that in every action of every force energy is expended, and that the sum total of this energy is constant. In the event, Faraday's notion of force remained as vague as that from which Helmholtz set out and from which he distilled the notion of energy.[51]

Helmholtz's achievement also depended on the progress of research in different fields to the point where a diversity of forces were recognized and thought to be conserved, and so in principle were capable of being seen in relation to a single entity—energy. In the fields of heat and electricity, research was stimulated by technology—both by the existence of steam engines in which the conversion of heat into mechanical energy could be investigated and by the demand for a more efficient prime mover than the engines then in use. Thus, in France, Sadi Carnot and Émile Clapeyron proposed theories to explain how mechanical energy was produced from heat in steam engines, and in England, James Prescott Joule conducted a series of experiments to find out if an electromagnetic engine might be more efficient. (His experiments, showing the interconvertibility of heat, electricity, and mechanical energy, were cited by Helmholtz.)

Helmholtz also was influenced by events in the Germany of his day. During the 1840s Germany passed through a series of social and political crises culminating in the abortive "liberal" revolution of 1848. The liberal reformers attacked not only the social and political arrangements of their time but also the dominant world view and ideology. Inherited from the Romantics, this world view represented both nature and society as organic entities in which the part, whether man or natural phenomenon, contributed to the maintenance of the whole. The reformers, on the other hand, demanding more freedom for the individual, took an atomistic and mechanistic view of nature and society, stressing the relative independence of the individual. This social and political crisis, and the movement of thought it generated, stimulated scientists to reexamine the foundations of their own disciplines and embrace the world view of mechanism (that natural processes are causally determined and capable of complete explanation by the laws of physics and chemistry).

This rethinking was intensified by the professionalizing of science. In their attempt to make science a recognized profession, scientists tried to become more socially responsible in the hope of winning public support. They not only thought more deeply about their science but, seeking to find some common ground with the educated public, they also were prone to adopt the rising world view. Nowhere was the process of rethinking more radical than in physiology, which was just emerging as an independent science, with its own professional organization, distinct from anatomy and medicine. Young physiologists, such as Helmholtz, enthusiastically embraced physical reductionism, the program of explaining biological processes by physical and chemical laws. They hoped to make their discipline a model for the older life sciences to follow. Reductionism in physiology and reconstructionism in politics stimulated Helmholtz to think boldly and to seek the common denominator of the forces of nature—energy.[52]

SUMMARY

The personality of the individual scientist influences his work far more than is generally realized. Different forms of research tend to attract different personality types. Theorists are apt to be adventurous, imaginative, and self-confident. Empiricists generally are careful workers, firmly committed to existing theories, and reluctant to invent hypotheses. Between them are scientists who propose hypotheses as well as gather data but do so within the horizons of established theories. Certain scientists involved in the discovery of the structure of DNA—Watson and Crick, Franklin and Griffith, and Wilkins— exemplify these three personality types.

Certain psychological factors enter into the scientist's research: personality traits, convictions, style of thought, subconscious thinking, unconscious motives, power of perception. Einstein, for instance, combined contraries in his nature, and believed in the ultimate harmony of the universe. Planck was conservative and believed that science should seek the absolute. Faraday thought pictorially and intuitively, but was deficient in mathematics. Maxwell combined physical insight with mathematical ability. Subconscious thinking plays an important role in research, especially in the formation of hypotheses. It appears that sometimes, after the scientist has gained a preliminary insight into a problem, his subconscious mind takes over and works out a solution for him. Both Darwin and Kekulé are known to have had this experience. The scientist also may be strongly influenced by unconscious motives, although we know little about this. Newton seems to have been especially driven by his unconscious. Scientists vary, too, in their powers of perception. Some, such as Carl Anderson (discoverer of the positron), are more original in their perceptions than others. Scientists, again, are subject to strong emotions. Heisenberg and Schrödinger, for instance, detested each other's ideas.[53]

However, the subjectivity of the individual scientist tends to be balanced by both the interpersonal criticism of the scientific community and the many different influences from the world around him. Such influences led Helmholtz to be the first to discover the law of the conservation of energy. I shall examine these influences at some length later.

NOTES

1. Emilio Segrè, *Enrico Fermi, Physicist* (Chicago: University of Chicago Press, 1970), p. 102.

2. At present we have few psychological studies of individual scientists, especially past scientists. Hence this chapter offers a very limited breakdown of certain aspects of the scientist's psyche.

3. On these three personality types, see Ian I. Mitroff, *The Subjective Side of Science*, pp. 84–87, 91–93, 284 n. 11. See also Ann Roe, "The Psychology of the Scientist," in *The New Scientist*, ed. Paul Obler and Herman Estrin (New York: Doubleday Anchor, 1962).

4. On Dalton see L. K. Nash, "The Origins of Dalton's Atomic Theory," *Isis* 47 (1956): 104–5; and J. R. Partington, "The Origins of the Atomic Theory," *Annals of Science* 4 (1939): 279. On Mendel see Gavin de Beer, "Mendel, Darwin, and Fisher," *Notes and Records of the Royal Society of London* 19 (1964): 192–225; L. C. Dunn, "Mendel, His Work and His Place in History," *Proceedings of the American Philosophical Society* 109 (1965): 189–98; and B. L. van der Waerden, "Mendel's Experiments," *Centaurus* 12 (1968): 275–88. On Ptolemy see Robert R. Newton, *The Crime of Claudius Ptolemy* (Baltimore: Johns Hopkins University Press, 1977).

However, if some scientists have been unscrupulous, others have been heroic. For instance, Joseph Goldberger, who identified the cause of pellagra as a dietary deficiency, injected both himself and his wife with vomitus and bodily secretions that had been thought to transmit the disease. Neither was affected.

5. Dorothy Michelson Livingston, *The Master of Light: A Biography of Albert A. Michelson* (New York: Scribner's, 1973), pp. 112, 334.

6. Russell McCormmach, "Editor's Foreword," in *Historical Studies in the Physical Sciences*, ed. Russell McCormmach, 3: xiii–xv.

7. P. L. Kapitza, "Recollections of Lord Rutherford," in *The Physicist's Conception of Nature*, ed. Jagdish Mehra (Dordrecht, Holland, and Boston: Reidel, 1973), pp. 761–62.

8. James D. Watson, *The Double Helix*.

9. Watson, like Wilkins, underestimated Franklin, thinking her a crusading feminist. As Ann Sayre has shown in her sympathetic study, *Rosalind Franklin and DNA* (New York: Norton, 1975), Franklin was given less than her due.

10. Peter Brian Medawar, *The Hope of Progress* (London: Methuen, 1972), p. 105.

11. Quoted by Philipp Frank, *Einstein: His Life and Times* (New York: Knopf, 1947), p. 49.

12. Quoted by Gerald Holton, *Thematic Origins of Scientific Thought*, p. 357.

13. *Ibid.*

14. See *ibid.*, pp. 357–67, and Russell McCormmach, "Editor's Foreword," in *Historical Studies in the Physical Sciences*, ed. McCormmach, 2:x–xi.

15. *Ideas and Opinions* (New York: Crown, 1954), pp. 36–40.

16. Quoted by Holton, *Thematic Origins of Scientific Thought*, p. 357.

17. Albert Einstein, "Autobiographical Notes," in *Albert Einstein: Philosopher-Scientist*, ed. Paul Arthur Schilpp, Library of Living Philosophers (La Salle, Ill.: Open Court, 1949), p. 9.

18. Quoted by Holton, *Thematic Origins of Scientific Thought*, pp. 236–37.

19. Carl Seelig, *Albert Einstein: A Documentary Biography*, trans. Mervyn Savill (London: Staples, 1956), p. 41. Quoted by Lewis S. Feuer, "Methods in the Sociology of Science: Rejoinder to Professor Agassi," *Philosophy of the Social Sciences* 6 (September 1976): 252.

20. Allie Vibert Douglas, *The Life of Arthur Stanley Eddington* (London: Nelson, 1956), p. 41. Quoted by Feuer, "Methods in the Sociology of Science," p. 252.

21. Quoted by Max J. Klein, "Max Born on His Vocation," *Science* 169 (1970): 361. Cited in Jeremy Bernstein, *Einstein* (New York: Viking, 1973), p. 192.

22. *Scientific Autobiography and Other Papers*, trans. Frank Gaynor (New York: Philosophical Library, 1949), p. 13. See Stanley Goldberg, "Max Planck's Philosophy of Nature and His Elaboration of the Special Theory of Relativity," in *Historical Studies in the Physical Sciences*, ed. McCormmach, 7:125–60.

23. *Scientific Autobiography*, pp. 13–14.

24. Max Planck, *A Survey of Physical Theory* (1925; reprint ed., New York: Dover, 1960), p. 166.

25. In a letter, quoted by Max Jammer, *The Conceptual Development of Quantum Mechanics* (New York: McGraw-Hill, 1966), p. 22.

26. *Scientific Autobiography*, pp. 45–46.

27. Goldberg, "Max Planck's Philosophy of Nature," p. 147.

28. On Faraday's view of the relation between theory (in this case his theory of gravitation) and experimentation, see William Berkson, *Fields of Force: The Development of a World View From Faraday to Einstein* (New York: Wiley, 1974), p. 116.

29. James Jeans, "James Clerk Maxwell's Method," in J. J. Thomson et al., *James Clerk Maxwell: A Commemoration Volume, 1831–1931* (Cambridge: Cambridge University Press, 1931), p. 92.

30. C. G. Knott, *The Life and Scientific Work of James Guthrie Tait* (Cambridge: Cambridge University Press, 1911), pp. 4, 5.

31. Lewis Campbell and William Garnett, *The Life of James Clerk Maxwell, with a*

Selection from His Correspondence and Occasional Writings (London: Macmillan, 1882), p. 175, n. 1.

32. Ibid., p. 133, n. 2.

33. Scientific Papers of James Clerk Maxwell, ed. W. D. Niven (Cambridge: Cambridge University Press, 1890), 1:155.

34. Scientific Papers of James Clerk Maxwell, 1:156.

35. A Treatise on Electricity and Magnetism (3d ed., 1891; reprint ed., New York: Dover, 1954), vol. 1, sec. 72.

36. Elementary Treatise on Electricity, ed. W. Garnett (Oxford: Oxford University Press, 1881), sec. 64. (I.e., the analogy is between the geometrical representations of heat conduction, fluid motion, and an electric current.) On Maxwell's way of considering Hamilton, see Mary Hesse, The Structure of Scientific Inference (Berkeley and Los Angeles: University of California Press, 1974), ch. 11.

37. E.g., the Russian physicist Peter Kapitza writes: "In science, in a certain phase of the development of new basic concepts, erudition is not the essential feature which enables a scientist to solve a problem, and . . . instead, the main thing here is imagination, concrete thinking, and, yet more importantly, daring. . . . The scientist's ability to solve major scientific problems of this kind, without demonstrating any clearcut logical structure in the process, is usually called intuition" (Peter Kapitza on Life and Science, Addresses and Essays Collected, Translated, and Annotated with an Introduction by Albert Parry (New York: Macmillan, 1968), p. 104.

38. Similarly in mathematics. Poincaré wrote of his own discoveries: "Most striking at first is the appearance of sudden illumination, a manifest sign of long unconscious prior work. The role of this unconscious work in mathematical invention appears to be incontestable." Quoted by Jacques Hadamard, The Psychology of Invention in the Mathematical Field (1945; reprinted, New York: Dover, 1954), p. 14.

39. Henri Poincaré, Science and Method, trans. F. Maitland (1914; reprint ed., New York: Dover, 1952), quoted by Mary Henle, "The Birth and Death of Ideas," in Contemporary Approaches to Creative Thinking, ed. Howard E. Gruber, Glenn Terrell, and Michael Wertheimer (New York: Atherton, 1962), p. 42. On creative thinking in science, see, in addition to Hadamard, Psychology of Invention, Arthur Koestler, The Act of Creation (New York: Macmillan, 1964), and Calvin W. Taylor and Frank W. Barron, eds., Scientific Creativity: Its Recognition and Development (New York: Wiley, 1963). On creativity in general, see my The Art and Science of Creativity (New York: Holt, Rinehart, and Winston, 1965).

40. Howard E. Gruber, Darwin on Man: A Psychological Study of Scientific Creativity, together with Darwin's early and unpublished notebooks transcribed and edited by Paul H. Barrett (New York: Dutton, 1974), chs. 6–8.

41. Quoted ibid., pp. 144, 159, 169.

42. In *To Be an Invalid: The Illness of Charles Darwin* (Chicago: University of Chicago Press, 1977), Ralph Colp, Jr., argues that Darwin became ill because he feared that he would be ridiculed and persecuted if he published his theory of evolution. In fact, Darwin was healthier when working on the evolution of species than he was at other times. [See William B. Provine in *Science* 196 (June 24, 1977), 1431–32.]

43. Frank E. Manuel, *A Portrait of Isaac Newton* (Cambridge: Harvard University Press, 1968).

44. Unfortunately Manuel provides only glimpses of these motives operating in Newton's research. He concentrates instead on the part they played in his relations with other scientists, as in Newton's unwillingness to tolerate rivals and his packing of the available scientific positions with his favorites during his long presidency of the Royal Society. Here unconscious motives operate within the profession rather than within the intellectual discipline of science. Nevertheless, Manuel's work should encourage other biographers to analyze in detail what he calls (*ibid.*, p. 64) "the intense personal character of Newton's struggle with nature" and to demonstrate that "the overwhelming fears, doubts, and insecurities of his early life . . . gave his science a particular style and on occasion tended to push him in one or another direction."

45. See Norwood Russell Hanson, *The Concept of the Positron*, ch. 9.

46. I have restricted my treatment of the psychology of the scientist to his work within the discipline of science. What aspects of the scientist's psychology are external to science? Anything in the psyche of the scientist that affects his research and his relations with his colleagues may be considered internal to science. The processes, however, by which any such psychical elements are acquired outside the professional life of the scientist are external to science. Thus Newton's need for security and his attitudes toward such men as Hooke and Halley are internal factors, whereas the departure of his mother at an early age and the death of his mother later—though not the effect of that death on his creativity—are external factors. (See chapters 9 and 10.)

47. Werner Heisenberg, *Physics and Beyond: Encounters and Conversations*, trans. Arnold J. Pomerans (New York: Harper & Row, 1971), pp. 73–75.

48. Quoted by Max Jammer, *Conceptual Development of Quantum Mechanics*, pp. 271–72. See Gerald Holton, *Thematic Origins of Scientific Thought*, pp. 132–33.

49. See Ian Mitroff: "It is important to emphasize that . . . commitments alone do not make for the objectivity of science. It is the presence of intense commitments coupled with experiments . . . , seemingly impersonal tests, arguments . . . , evidence, and general paradigms that make for the objectivity of science. No single one of the elements does the job. Science, as opposed to other systems of knowledge, is distinguished by the fact that, if not in theory then in actual practice, it has learned

how to make use of strong determinants of rationality (testing, evidence, etc.) plus strong emotional commitments." *Subjective Side of Science*, p. 249.

50. Yehuda Elkana, "Helmholtz's 'Kraft': An Illustration of Concepts in Flux," in *Historical Studies in the Physical Sciences*, ed. McCormmach, 2:263–98; and *The Discovery of the Conservation of Energy* (London: Hutchinson International, 1974). Elkana's book is reviewed critically by Peter Clark, "Elkana on Helmholtz and the Conservation of Energy," *British Journal for the Philosophy of Science* 27 (June 1976): 165–76. For a different view than Elkana's see Thomas S. Kuhn, "Energy Conservation as an Example of Simultaneous Discovery," in *Critical Problems in the History of Science*, ed. Marshall Claghett (Madison: University of Wisconsin Press, 1959), pp. 321–56. Kuhn argues that the law of the conservation of energy was "discovered simultaneously" by at least a dozen scientists in the 1010s. For yet another view, see P. M. Heimann, "Helmholtz and Kant. The Metaphysical Foundations of *Über die Erhaltung der Kraft*," *Studies in History and Philosophy of Science* 5 (November 1974): 205–38.

51. Elkana, *Discovery of the Conservation of Energy*, pp. 130–32.

52. See Everett Mendelsohn, "Revolution and Reduction: The Sociology of Methodological and Philosophical Concerns in Nineteenth-Century Biology," in *The Interaction between Science and Philosophy*, ed. Yuhuda Elkana (Atlantic Highlands, N.J.: Humanities Press, 1974), pp. 407–26.

53. Many who have studied the lives of great scientists might claim that some of them are a greater challenge than the ones I have selected. I regret that because of space limitations I have had to overlook such persons. Still, I cannot resist citing one more. Louis Pasteur was a martinet who brooked no interference with his work. He was imperious and stubborn. Irked by Pasteur's highhandedness, one of his assistants secretly stopped vaccinating chickens and let the vaccine stand for nearly three weeks before resuming. The chickens did not die as expected, because the vaccine had become too weak from exposure. Pasteur then realized for the first time that immunization could be built up by injecting weak vaccines followed by stronger ones. The chickens still lived when given the full dose after two previous ones, and the lives of countless chickens have been saved ever since. This is another dramatic instance of discovering something truly remarkable as a result of a peculiar personal relationship, a freak accident, or a deliberate misdeed. Although Pasteur was known as an extravagant theorizer with an unusual imagination, he was sure that "chance favors only the mind that is prepared." Pasteur's mind indeed was prepared.

189

Chapter 8

The Scientific Community

No single scientist can discover all there is to be known about nature. Even if he could, his findings would be of no value to others unless they could be verified. Thus science necessarily is a social enterprise. The individual scientist depends on and contributes to a professional community. What is this community like?

The scientific community is an association of persons bound together neither by law nor chain of command but by the communication of information—through journals, conferences, informal discussion, and other channels. Communication is coordinated by institutions such as specialist societies and invisible colleges. By means of these institutions and channels, and the reward mechanism operating through them, the scientific community seeks to attain certain ends that contribute to the overall aim of extending knowledge of nature. These ends are to maintain research standards, reconcile the interests of the individual scientist and the scientific enterprise, promote competition and cooperation, and stimulate innovation.

INSTITUTIONS

Scientific societies create and maintain channels of communication in which claims to knowledge are made and assessed. Each society organizes research in a discipline (e.g., physics) or specialty (e.g., particle physics, astrophysics). Through its channels innovating groups, communicating informally within themselves, bring competing research programs to the attention of members. *Professional associations* control entry to societies, promote professional ethics, and represent the profession to the public. Some societies, especially those organizing research in disciplines (e.g., the American Chemical Society), carry out both research and professional functions.

The chief innovating vehicle in science is the research tradition. Nowadays at least, the institution which carries the research tradition is not as a rule the specialist society but the invisible college.[1]

Invisible Colleges. An invisible college is a group or school of about ten to a hundred scientists working either within or as a single research tradition. Its members keep in close touch, usually by word of mouth, and avoid the slower channels of formal communication. The group may be one of several applying a comprehensive research program to different classes of phenomena and problems, as in Kuhnian normal science. Or it may be one of several traditions competing within a specialty, as in the case of the Bohr, Rutherford, and Fermi groups in nuclear physics. Or it may be consciously revolutionary, launching a new research tradition against an established one. Between 1947 and 1958, for instance, Max Delbrück's phage group[2] developed the new tradition of molecular biology in opposition to the specialty of biochemistry. The group usually does most of its work in one or a very few places. Rutherford's group worked at Manchester and then Cambridge, Bohr's at Copenhagen, and Fermi's in Rome. The phage group spent their winters at Cal Tech and their summers at Cold Spring Harbor, New York.

An invisible college often is inspired by an outstanding scientist who formulates its basic assumptions, makes public statements on its activities, and evaluates the work of its members. Delbrück, for instance, was a vigorous leader, strongly pushing some projects and equally strongly suppressing others. He set priorities for research and standardized techniques and models.

Rutherford, equally vigorous, directed less. He particularly valued independent thought. Having given a man a problem, he would leave him on his own. Researchers were free to design their own experiments and run equipment as they wished. But he insisted on hard work. As one of his colleagues said, "He drove us mercilessly, but we loved him for it."[3] He would come around twice a day, pausing briefly to boom, "Why don't you get a move on? When are you going to get some results?"[4] He was jovial but easily irritated by delays. To a research student from New Zealand, building an elaborate apparatus, he said genially but firmly: "Well, sir, still making and breaking? When are you going to do some physics?"[5] Urging people to get on with it seems to have been reflex with him. At Cambridge during a building strike he even told a volunteer bricklayer to get a move on.[6]

An invisible college carries out a research program. This usually consists of a conception of the central problem or problems to be solved, a set of assumptions from which solutions are to be derived, and techniques and models to be used in finding solutions. The phage group's research program was set out by Delbrück in his historical address to the Connecticut Academy of Sciences (1948), "A Physicist Looks at Biology."[7] Its goal was to discover how biological information is transmitted from one generation to another.

Rutherford's men, on the other hand, distrusted abstruse mathematical theories. Theorists, Rutherford once said, "play games with their symbols, but we, at the Cavendish, turn out the real solid facts of nature."[8] When questioning theorists, he often would say, "I am a simple man and I want a simple answer." Proposing a vote of thanks to Heisenberg, he declared: "We are all much obliged for your exposition of a lot of interesting nonsense."[9] Nevertheless, he abandoned his prejudices whenever one of these theories supported his own experiments. On one such occasion he is reported to have said, "There's more in old Planck than meets the eye."[10] His research program was experimental rather than theoretical: to discover the structure of the atom, assuming it to be planetary, by observing particle collisions—or as Rutherford used to say, "Smash the atom!"[11]

A group with a radically new program must win support gradually by advertising its results and actively recruiting members.[12] It cannot do so effectively unless the morale of its members is high. The original members tend to be especially dedicated and creative. Their camaraderie and commitment often are expressed in a distinctive research- and life-style. This statement by a member of the phage group gives an idea of the group's morale:

Once started with bacteriophages, I continued research on them because it seemed that here one could do more interesting experiments
than with any other material. Furthermore, the engaging group of
people working on these viruses was highly compatible; among ourselves we saw, and still see, very little competitive secrecy and backbit-
ing; we were all interested in cooperating with each other to promote
the work as a whole.[13]

Similarly the Russian physicist Peter Kapitza described his 13 years at
the Cavendish Laboratory with Rutherford as the happiest of his career. Calling Rutherford a "great and remarkable man, Kapitza said
he also was "undoubtedly very generous, and I think this is one of the
main secrets which explains why so many first-class scientists came
from his laboratory. There was always an atmosphere of freedom and
efficiency there."[14]

A leading feature of the phage group's research style was "the
principle of limited sloppiness," meaning roughly that an experiment
was likely to have more interesting results if it was not done too precisely.[15] Delbrück was not experimentally minded. He urged phage
workers to value a theory more highly than data and to spend at least
one day a week away from the laboratory just thinking. He also urged
publishing a few excellent papers rather than many marginal ones. A
characteristic feature of the group was the blunt intellectual honesty
evident in seminars and conversations as well as in papers.

A group life-style may include mock ceremonies,[16] citations,
awards, and sometimes a sport (camping for the phage group, moun-
taineering and table tennis for the Copenhagen school). A former
phage worker recalls the group's camping mania:

Sometimes Delbrück would proclaim Wednesday and Thursday as a
weekend, to avoid crowds and highway traffic on camping trips. The
first camping trip in which I participated, into the Anza desert, was
fairly typical. We just kept driving until the car got stuck in the sand,
and that determined the campsite. Most of the following day was spent
in digging the car out. Such visits to the desert were (and still are) the
favorite means of entertaining visitors, although some people, like
Luria, will not come to Cal Tech unless guaranteed immunity from

193

camping. It was on one such trip in 1950 that André Lwoff invited me to spend a year in his laboratory at the Institut Pasteur in Paris.[17]

Another aspect of the phage group's life-style was the writing expedition. Delbrück would take several students to Cal Tech's Marine Biology Station at Corona Del Mar and lock them up until each had written a report on his work. Mrs. Delbrück would type the reports, and the participants then would criticize one another's papers. These sessions stimulated rewriting, and everyone managed to complete at least one paper.

An invisible college usually is short-lived—10 to 15 years on an average. As the group grows, its ties weaken. When it is recognized as an established specialist community, it loses its sense of identity and settles down as a conventional research tradition.

The Wider Scientific Community. In this book I am concerned mainly with the basic research community. But the scientific community as a whole is much wider. It includes applied scientists, engineers, technicians (as in medicine and public health), administrators, and people in educational institutions. In 1975 the American scientific community numbered about a quarter of a million people, of whom roughly 40 percent were in industry, 30 percent in education, and 20 percent in government. The remaining 10 percent worked for nonprofit organizations. Basic research was done by less than a fifth of all American scientists.

Nonacademic institutions usually are not interested in the advance of knowledge for its own sake. Some government and nonprofit laboratories, and the laboratories of some of the larger companies (such as Bell Telephone, IBM, and General Electric), support a measure of basic research, but they do so partly to attract talented scientists and partly on the offchance that it may be profitable. True, Bell Labs has had some striking successes. In 1925 Clinton Davisson and Lester Germer discovered that the electron had wave properties. In the early 1930s on a homemade antenna Karl Jansky picked up cosmic microwave background radiation—radiation remaining from the Big Bang itself—but did not realize its significance.

194

In 1965 Arno Penzias and Robert Wilson came across it again and identified it. In 1948 John Bardeen, Walter Brattain, and William Shockley invented the transistor. But these are the exceptions. Most government and industrial research is undertaken for a specific practical end. Management generally chooses research topics and does not let the scientist pursue interesting sidelines.

Journals. The most important form of scientific communication is the paper published in a professional journal. A scientific paper is written in an impersonal style and observes many conventions. The scientist is expected to give due credit to previous work on the problem, to refrain from unnecessary criticism of other scientists, and to claim no more than he can support with evidence. The paper then is read and checked by other specialists. Sometimes it will set off a succession of other papers on the same problem, each correcting the errors of its predecessors and proposing a fresh solution. This graduated process of criticism—the second research cycle—keeps incompetent work to a minimum. It means that every fresh claim to scientific knowledge must be presented in such a way that it is acceptable to specialists working in an established research tradition (or, if it is submitted to a breakaway journal, then to specialists who apply with equal vigor the standards of a competing tradition).

Well over a million papers are published each year in more than 35,000 journals. One might think that, with so many journals, anything can be published somewhere—and swiftly. But delays are notoriously long (often up to a year) and journals differ enormously in prestige.[18] A recent study found that of 1,842 current British science journals only 590 were cited, and that in the 1965 Science Citation Index 95.2 percent of the citations were to only 165 periodicals (15 percent to *Nature*, and 51 percent to a mere 11 periodicals).[19] If citation measures use, if use indicates value, and if British scientific journals typify scientific literature the world over, then nearly all contributions of real value are published in about 10 percent of the world's scientific periodicals. Turning to a particular discipline, we find that in 1968 77 percent of American physicists read *Physical Review* consistently; 59 percent read *Physical Review Letters*

195

frequently; and no more than 25 percent read any other journal frequently.[20]

Conferences. Scientific conferences range from international congresses to meetings of specialist societies, and from open sessions of the American Association for the Advancement of Science to "invitation only" seminars. As a rule scientists attend them more to meet other scientists than to listen to scheduled speakers. Current research problems are canvased most intensely in informal discussions between speeches, and in conversations over lunch and in the corridors.

Among the most famous are the Solvay conferences in physics held every three years in Brussels.[21] These were the brainchild of the German physical chemist Walter Nernst. A natural entrepreneur, Nernst had invented an electric incandescent lamp and had sold the patent to William Siemens for a small fortune. In the spring of 1910 he met the Belgian industrial chemist and amateur scientist, Ernest Solvay, in Brussels. Solvay asked Nernst if there were any way he could submit his scientific ideas to leading physicists like Planck and Poincaré. He also mentioned that he was interested in the crisis created for classical physics by relativity and quantum theory. Nernst at once proposed an international conference on the growing conflict between quantum theory and statistical mechanics. Solvay accepted, and the conference was convened.

Twenty-one leading physicists met from October 30 to November 3, 1911. At first only a few of them—such as Planck, Einstein, Lorentz, and Wien—knew or cared much about quanta. No one mentioned Einstein's photon theory of 1905 or Paul Ehrenfest's paper, published a few days before the conference, showing that a quantum law was both a necessary and a sufficient condition for the solution of the black-body problem. The turning point in the conference was the conversion of the leading mathematical physicist in the classical tradition, Henri Poincaré, during his first encounter with quantum theory. Barely a month later he proved mathematically that Planck's law was the only possible solution to the black-body problem, and that the principle of the quantization of energy was both a necessary and a sufficient condition for the truth of that law. Poincaré's paper was the wedge which split the almost unanimous opposition of physicists to the quantum theory. Even James Jeans, unpersuaded at the conference, later surrendered.

Einstein, meeting many of his colleagues as an equal for the first time, made a deep impression on them. Frederick Lindemann, later Lord Cherwell and scientific adviser to Winston Churchill, wrote to his father, "I got on very well with Einstein who made the most impression on me except perhaps Lorentz. . . . He says he knows very little mathematics, but he seems to have had a great success with them."[22] Almost half a century later he wrote, "I well remember my co-secretary, M. de Broglie, saying that of all those present Einstein and Poincaré moved in a class by themselves."[23] Poincaré also was struck: "The future will show more and more the worth of Mr. Einstein, and the university intelligent enough to attract this young master is certain to reap great honor."[24] The university turned out to be Berlin, to which Einstein was invited by the Prussian Academy of Sciences on the recommendation of Planck and Nernst. However, Einstein himself seems to have been in two minds about the conference. Although he wrote appreciatively to Solvay, he confided to his friend Michèle Besso that the meeting was a "witches' sabbath." "I did not benefit much," he complained, "as I did not hear anything which was not known to me already."[25]

Subsequent Solvay conferences considered the structure of matter, atomic nuclei, elementary particles, astrophysics, and so on. One of the most exciting was the fifth conference (1927), which confronted the new quantum mechanics complete with the Copenhagen interpretation. Hendrik Lorentz, in the chair, tried to give the floor to one speaker at a time. But feeling ran so high that several participants often spoke simultaneously, each in his own language. Ehrenfest went to the blackboard and wrote, "The Lord did then confound the language of all the Earth!" During a lecture he passed a note to Einstein, saying, "Don't laugh! There is a special section in purgatory for professors of quantum theory, where they will be obliged to listen to lectures on classical physics ten hours every day." Einstein replied, "I laugh only at their naïveté. Who knows who will have the laugh in a few years!"[26]

197

Informal Contacts. Because of publication delays and the difficulty of keeping up with the literature, scientists rely more and more on informal communication. The average scientist knows a great many researchers, though he may communicate regularly with only a few of them.[27] Many circles of acquaintances overlap; hence loose, temporary networks readily are formed. When a network becomes an invisible college, communication intensifies. James Watson's *Double Helix* gives an inside view of this side of an invisible college. At Cambridge, Watson and Crick communicated and received hot data and ideas by personal channels. They extracted crucial information from colleagues and visitors, such as biochemist Erwin Chargaff and structural chemist Jerry Donahue. Watson attended major conferences at Paris and Royaumont, haunted King's College, London, for news of a competing group there, and got an advance copy of Linus Pauling's latest paper from his son Peter at Cambridge. Journals were used chiefly to read up on basic knowledge and secure priority for claims.

MAINTAINING RESEARCH STANDARDS

Education. A prime concern of the scientific community is to establish and maintain standards of research. A disposition to obey these standards is instilled in the scientist by his education and, as we shall see, is maintained thereafter by various social mechanisms. The scientist's education acquaints him with a body of established knowledge and techniques. Some of this information is tacit and informal, and can only be learned from experience and by emulating qualified practitioners.[28] Formal scientific knowledge is highly organized and normally is presented more rigidly than the content of other disciplines. A stricter intellectual conformity is imposed on the student. When he is not absorbing basic knowledge, he is seeking the correct answers to precise problems posed from the standpoint of an established research tradition. Both knowledge and problems are presented by textbooks that may differ in depth and detail but not usually in intellectual approach.

198

Scientific education is a powerful molding force for other reasons. It takes up most of a student's time and energy. It makes science students more dependent on their teachers for technical help and eventually for employment. Recruitment for it is highly selective. Students are examined regularly, and those who cannot understand or will not accept the accredited body of knowledge are quickly excluded. All in all, as Thomas Kuhn has said (somewhat wryly), nothing could be better calculated to produce "mental sets" for fitting nature into prescribed boxes than the present method of scientific education.[29]

This view of science education as a molding force is supported by Liam Hudson's studies of "convergers" and "divergers" among English boys of high-school age. Convergers scored well on questions that had one correct answer but poorly on questions that could be answered from various points of view. Divergers, on the other hand, did rather badly on highly structured tests but correspondingly well on tests requiring them to think "fluidly" and "flexibly." Hudson found that between three and four divergers went into the humanities for every one choosing the sciences, and conversely that between three and four convergers entered the sciences for every one who selected the humanities.[30] This certainly is not to say that a science student cannot think originally or creatively but rather that he is taught to do so within the limits of accepted knowledge.

Specialization limits the scientist's thinking. Francis Crick, for example, working with James Watson on the structure of the DNA molecule, did not know Erwin Chargaff's famous rules for the ratios between the bases in this molecule. Chargaff and Crick were both biochemists. Chargaff was a world leader in nucleic-acid chemistry. But Crick was a protein chemist and X-ray crystallographer. He had come to DNA by accident while writing his dissertation. Chargaff met Watson and Crick while visiting Cambridge University, and was not impressed by them. Watson describes the meeting:

> The high point in Chargaff's scorn came when he led Francis into admitting that he did not remember the chemical differences among the four bases. The faux pas slipped out when Francis mentioned Griffith's calculations. Not remembering which of the bases had amino

groups, he could not qualitatively describe the quantum-mechanical arguments until he asked Chargaff to write out their formulas. Francis's subsequent retort that hc could always look them up got nowhere in persuading Chargaff that we knew where we were going or how to get there.[31]

Access to Research Facilities. Once graduated, a scientist seeks a job and research facilities. For the first he needs the recommendation of his professor, for the second the support of a senior scientist in the field. Consider Augustin Fresnel in 1815, a civil engineer, self-taught in physics, eager to break into the Parisian scientific establishment dominated by the school of Pierre Simon Laplace and Claude Louis Berthollet.[32] Fresnel was interested in the wave theory of light, to which the school was opposed. So was Dominique François Arago, a member of the school. Arago had fallen out with another member, Jean-Baptiste Biot, a notorious "claim-jumper," whom he had accused of preempting one of his own discoveries. Having abandoned the corpuscular theory of light, Arago met Fresnel socially in the summer of 1815 and advised him to read the English wave theorist, Thomas Young. When Fresnel sent his first paper on diffraction to the Institut Français, Arago interceded for him. Most papers submitted to the Institut by nonmembers were read not by the body as a whole but by a reporter, who could kill them if he chose. Had Fresnel's paper been assigned to Biot, the expert in optics, it might have been ignored or damned with faint praise. Arago, however, saw Fresnel as an ally in the fight for the wave theory. He got in touch with the head of the corps of engineers and arranged for Fresnel to come to Paris and continue his research. There the two men collaborated on experiments which strengthened the wave theory. In March 1816 Arago not only read a glowing report of Fresnel's work before the Institut but also published Fresnel's paper on diffraction in a journal which he co-edited. A year later he got him a permanent position in Paris. As a recent historian has put it, "The acceptance of the wave theory in France owes as much to Arago's support as it does to Fresnel's experimental insight."[33]

Or take Fermi.[34] The Italian physicist knew from the start that

200

he had to make his way in an old-fashioned, undistinguished physics community. Soon after receiving his doctoral degree in 1922, he cultivated the acquaintance of some of Italy's most distinguished mathematicians, hoping thus to succeed in a weak academic field with support from leaders of a strong one. While looking for a job, he wrote as many papers as he could, seeking to impress his judges by sheer volume of publication. On graduation he presented himself to Italy's leading physicist and statesman of science, Orso Mario Corbino, a former cabinet minister. Through Corbino's efforts a chair in theoretical physics was created at the University of Rome, and Fermi was appointed to it. Again through Corbino's influence, Fermi was elected at age 28 to the Royal Academy of Italy, the only physicist member. Similarly, Corbino saw to it that everyone in Fermi's circle got appointments and funds. He even recommended Fermi to students at a time when almost no one at the University of Rome studied physics. Without Corbino's protection, Fermi and his research group hardly would have survived the opposition of established scientists, notably the other Rome physics professor.

Exercise of Authority. In science, authority usually is exercised informally. Irritated by Crick's delay with his dissertation, Lawrence Bragg, head of the Cavendish Laboratory, told Crick and Watson to stop work on DNA. The two said nothing, but carried on secretly. A rare instance of the direct exercise of formal authority was the decision of the Paris Académie des Sciences in favor of Louis Pasteur's theory that airborne germs cause fermentation and putrefaction, as against Félix Pouchet's theory that germs are generated spontaneously. Noting that an infusion of yeast became putrid less often at high altitudes where germs were scarce, Pasteur argued that materials putresce when germs enter them from the air. Noting, on the other hand, that an infusion of hay contained living germs even at high altitudes, Pouchet claimed that germs appear spontaneously. At Pasteur's request the Académie des Sciences formed a commission to pass judgment. Pouchet rejected the commission's terms and withdrew from the contest. The Académie then decided for Pasteur.

201

Advice of Colleagues. A scientist also may be warned informally by his colleagues. Pasteur was warned several times. His first venture into crystallography was well received, because he did original work within an established tradition. But then he started to experiment and generalize extravagantly in the area of molecular and crystalline dissymmetry. He proposed, for instance, that life was an effect of the dissymmetry of the universe, and he used a clockwork mechanism to keep plants rotating in different directions. Many of Pasteur's techniques and conclusions struck his fellow scientists as absurd, and to a man they urged him to give them up. Getting nowhere, Pasteur turned to the problem of fermentation, a process thought to result from purely chemical reactions. He then proposed his germ theory. At once he was attacked by such leading chemists as Justus von Liebig, Jakob Berzelius, and Friedrich Wöhler who, 20 years before, had crushed an earlier vitalist theory that fermentation was produced by the organic growth of yeast cells. Wöhler had even written a skit, which Liebig had embellished and published in his *Annalen der Chemie*. (Yeast was described as consisting of eggs that turned into animalcules shaped like distilling apparatus. These lived on sugar, digesting it into carbonic acid and alcohol, and excreting them separately—as could be seen clearly under a microscope!) For all his skill in experimentation and debate, it took Pasteur years to win a majority of scientists to his side, and many of his opponents carried their views to their graves.

Access to Periodicals. To establish the priority and value of his work, the scientist must publish it in a professional journal. Who decides what goes into these journals? Mostly referees, who judge whether the claims made in a paper are of interest, whether the reasoning is sound and the evidence adequate, whether sufficient credit is given to other scientists in the field, whether the style is clear—in short, whether the paper is up to standard. Generally two or more referees are consulted, and a paper usually is rejected only if both strongly recommend against it. A journal selects its referees from scientists throughout the specialty, so that work rarely is submitted to the verdict of a hostile clique. Also, many referees are reluctant

202

to advise rejection; they remember some famous manuscripts that were turned down. For example, the leading British journal, *Nature*, rejected—ostensibly for lack of space—the first paper by Hans Krebs on what came to be called the "Krebs cycle," a theory of living processes in the cell and now one of the foundations of modern biochemistry.

Editors may judge papers without consulting referees. Between 1948 and 1956 the two editors of the *Physical Review* personally assessed nine out of ten papers submitted by high-ranking physicists but only six out of ten submitted by physicists of the lowest rank. Other papers went out to referees who were expert in the areas concerned." The *Physical Review* now consults over 1,000 referees a year. In one specialty, theoretical high-energy physics, 30 percent of all members had done some refereeing.[36] Control over publication is divided, then, between the editor (in the case of a general periodical, representing the discipline) and the referees (representing the specialties). The editor retains ultimate control, however, and may deny space to members of a rebellious group, who sometimes then establish their own journal.

One of the greatest editors of all time, Johann Christian Poggendorff, founded the celebrated *Annalen der Physik und Chemie* in 1824 and supervised it for more than half a century. Broadminded and altruistic, he welcomed to its pages some of the pathbreaking scientists of his day, such as Faraday (whom he published 76 times), Wöhler (65 times), and Liebig (56 times). Nevertheless, Poggendorff's sympathies had their limits. Trained as an apothecary's apprentice, he shared some of the hostility of his generation to *Naturphilosophie* (the theory that nature is a manifestation of an evolving World Spirit) and to all speculation not closely supported by experimental facts. Thus he turned down early papers on the conservation of energy by Julius Robert Mayer and Hermann von Helmholtz and also a paper by Ernst Mach. Mayer revised his paper, which contained serious errors, and published it the following year in another journal. Helmholtz, on Poggendorff's advice, took his paper to an independent printer. Mach, however, who already had published several times, was so discouraged that he delayed publishing his work on inertia. "If I ran up against the physics of the schools in

203

so simple and clear a matter," he wrote later, "what could I expect in a more difficult question?"[37] Poggendorff, who had found Helmholtz speculative, no doubt was horrified by Mach's bold thesis that the law of the conservation of energy can be reduced to a logical consequence of the law of causality.

THE REWARD MECHANISM

Research standards are maintained chiefly through the reward mechanism. This mechanism also serves the community's other purposes: it reconciles the interests of the individual scientist and the scientific enterprise, fosters competition and consensus, and promotes innovation. It thus is the most important element in the activities of the scientific community.

The reward mechanism operates on the principle that the scientist submits his work to his peers for their use and receives in return their professional recognition. Recognition takes many forms, ranging from the highly informal (such as requests for advice and invitations to address seminars) to the highly formal (such as the Nobel Prize or membership in honorary societies). Perhaps the greatest tribute that scientists can pay to one of their number is to work in his research tradition. Indeed, to have founded a research tradition is the supreme achievement of any scientist. Another coveted reward is "eponymy," or the attaching of a scientist's name to his achievement, as in Halley's comet, Brownian movement, Darwinian evolution, Planck's constant, and the Doppler effect. Yet another is the frequent citation of one's papers by other scientists.

Is the Reward Mechanism Fair? Desire for professional recognition is not the only motive for doing research. Many scientists claim that they solve problems for sheer intellectual pleasure. But it is the chief institutionalized motive and therefore should operate fairly. Does it?

No, say some observers, who argue that well-known scientists tend to be rewarded more than others. Sociologist Robert K. Merton

204

has called this "the Matthew Effect" ("For unto everyone that hath shall be given, and he shall have abundance: but from him that hath not shall be taken away even that which he hath," Matthew 25, xxix).[38] The history of science reveals many classic cases. John James Waterston's masterly paper on the kinetic theory of gases, which anticipated Joule's work by 20 years, was rejected in 1845 by a referee for the Royal Society who called it "nothing but nonsense, unfit even for reading before the Society." Waterston abandoned science in despair. Fifty years later Lord Rayleigh, having discovered Waterston's paper in the library at Burlington House, immediately proclaimed its historic importance and eulogized Waterston as a man of "marvelous courage." Mendel's papers on heredity, published in a little-known journal, were ignored by his contemporaries. Deeply hurt, Mendel refused to publish his later work, which now is lost. On becoming abbot, he gave up research altogether. Jean Baptiste Fourier waited 13 years for the French Académie des Sciences to publish his classic theory on the propagation of heat. There also is a telltale case in which a verdict on a paper was reversed when the name of its eminent author became known. Lord Rayleigh's name did not appear on a paper he submitted to the British Association for the Advancement of Science, which rejected it. When the Association learned who the author was, the paper was found to be worth publishing after all.

The Matthew effect is thought by many to operate today. Nobel laureates benefit from it when they are elected to the National Academy of Sciences. In the case of joint papers and simultaneous discoveries, the best-known name tends to be remembered. It generally is the already successful who are invited to accept honorary degrees, sit on policy committees, and contribute papers.

Nevertheless, a recent study has found little evidence of a Matthew effect, at any rate in the American physics community.[39] According to Jonathan and Stephen Cole, physicists are rewarded according to merit. The quality, and not the quantity, of a physicist's work usually determines whether he is honored or promoted. The author of a few good papers is not passed over for a prolific publisher of trifles. The papers of eminent scientists are not, as a rule, cited more readily than papers by others.

205

Does the advance of science depend on the work of a few talented scientists or on the contributions of the rank and file? The Coles reply that since most published papers are cited only once or not at all, science would lose little or nothing if the vast majority of papers now being published never were written. The Coles may be right. Unfortunately, they rely on citation counts as evidence of the quality of a scientist's work, notwithstanding that in the invisible colleges much influence is exercised by word of mouth. Until that influence has been taken into account, one cannot use their methods to justify the claim that most scientists contribute little, if anything, to the growth of science.[40]

But who judges the intellectual value of these contributions? The most respected judges are those who have made outstanding contributions to knowledge according to current standards. These men occupy positions of power in the scientific community; they distribute recognition; they know what research needs to be done within the established tradition, since they have guided and may even have created it. Professional authority and intellectual achievement for the most part coincide. Hence the scientific community, whose purpose is to promote research, is controlled by those who have done the best research (again, judged by current standards) rather than by a corps of administrators with interests of their own.

Does this mean that science is controlled by its disciplinary elites? Not entirely. As we have seen, the quality of scientific work is controlled mainly through hundreds of professional publications of every description. Papers are judged partly by editors but even more by referees competent in problem areas. By deciding for or against a particular manuscript, and by having his assessment passed on to the author, each referee plays his part in maintaining research standards. Thus rank-and-file scientists can influence decisions affecting the course of research, even if they cannot make them. Although the scientific community is stratified by ability, it is not a purely elitist community. Insofar as the power to influence is distributed widely through the specialties, it also is a participatory one.[41]

Nevertheless, the reward mechanism operates fairly and rationally only within limits, since the judges of last resort—the editors and the most respected referees—are the established scientists. Work which challenges an established tradition often is resisted and sometimes ignored. Copernicus, Galileo, Thomas Young, Fourier, Waterston, Mendel, Planck, Wegener, and many others were criticized or spurned by the scientific establishment. How are we to cope with this state of affairs? Principally, I suggest, through discussion of alternative metaphysical blueprints (see chapter 4) well before present theories run into serious difficulties. In this way radically new ideas can be anticipated, and so treated with respect when they finally are published as testable hypotheses. The established scientists will still have the last word but they will be made more tolerant.

COMPETITION

The reward mechanism intensifies competition between scientists and thus speeds up the growth of science. Scientists strive to be the first to make discoveries. Passionately, even deviously, they defend their claims to priority. Galileo first announced some of his telescopic discoveries in the form of tortuous anagrams, so as to protect his priority without revealing what he had observed.[42] When the same discovery is made more or less simultaneously by two or more scientists, a dispute often arises as to who made it first. The history of science is full of long, bitter wrangles over priority: Newton's wars with Hooke over problems in optics and celestial mechanics, and with Leibniz over the invention of the calculus; the three-cornered contest between Henry Cavendish, James Watt, and Antoine Lavoisier over who was first to prove the compound nature of water; the battle between John Couch Adams and Urbain LeVerrier over who had first deduced the existence and predicted the position of the planet Neptune.[43] When the claimants are unwilling to argue the matter themselves, their supporters argue it for them, as in the debate over priority in the discovery of the element argon by William Ramsay and Lord Rayleigh. Yet the award of priority to one scientist does not necessarily diminish the intellectual achievement of others who independently made the same discovery later; nor does it reduce the intellectual pleasure they had from solving a difficult problem. They seek priority for the sake of recognition, because only the first discoverer usually is recognized. This follows from the hard fact that the scientific community sees no value in having a particular contribution repeated.

Even when his research is fairly routine, the scientist is apt to fear that his findings may be anticipated. When the stakes are high, his fears intensify. We may well sympathize with Watson and Crick when Peter Pauling stood before them, his father's manuscript sticking out of his pocket. Had Pauling really discovered the structure of DNA? Had they been scooped? Watson recalled how his "stomach sank in apprehension on learning that all was lost." He grabbed the manuscript from Peter's pocket and read it. Linus had blundered! "The blooper was too unbelievable to keep secret for more than a few minutes." Watson and Crick spread the news:

> Then . . . Francis and I went over to the Eagle. The moment its doors opened for the evening we were there to drink a toast to the Pauling failure. Instead of sherry, I let Francis buy me a whiskey. Though the odds still appeared against us, Linus had not yet won his Nobel.[44]

In the rapid growth of physics during the 1950s, the race for priority became so fierce that many physicists began reporting their findings in short letters to the bi-monthly *Physical Review*. Before long the periodical was so swamped that a new journal, *Physical Review Letters*, was created, "devoted entirely to the fastest possible publication of short notes on important discoveries." Within a year the editor complained that he was getting too many contributions too insignificant to be worth publishing even as notes, and that some scientists were trying to outstrip their competitors by announcing their results to the daily press.[45]

Does the growth of science demand intense competition for professional recognition, or could the scientist be rewarded in some other way? One might argue that competition diminishes scientists by arousing hostility, and that it diminishes science by replacing intellectual concern with personal anxiety. One might even argue that competition is economically inefficient to the extent that time is wasted in priority disputes, and energy and resources are duplicated as several scientists race one another to the solution of a problem that could have been left to one of them. Surely the scientific enterprise would survive an end to competition.

Why, then, should scientists not do research from the motive which most of them actually cite, the satisfaction of advancing natural knowledge? Why should not all solutions of equal merit receive equal recognition? After all, in other times and cultures—in medieval Europe, classical Greece, and Hellenic Alexandria—science was done mostly from love of learning.

Yet ever since the Renaissance Western societies have believed in competition for possessions and prestige. The scientific community cannot seal itself off from the ideology of the society in which it exists. If other professionals compete fiercely for influence and affluence, how can we expect scientists not to? Why should we think that scientists are more altruistic? Science will remain competitive as long as society does. For science is affected more strongly by its sociocultural setting than most people realize. Nor is competition necessarily harmful. On the contrary, it stimulates people to develop their abilities to the utmost and to that extent to realize their humanity. Since we must live with competition (for no civilized society lives without it), we must resolve to check its excesses rather than hope to abandon it.

SUMMARY

The scientific community is an association of persons bound together by the communication of information. Communication is coordinated, and research is promoted, by scientific societies, professional associations, and invisible colleges. Scientific societies establish and maintain channels of communication. Professional associations regulate entry to societies, maintain ethics, and represent the societies to the outside world. Invisible colleges, informally organized and usually short-lived, do much of the original research in science and often launch new research programs. They frequently have their own research style and sometimes their own life-style.

The main channels of formal communication are journals and

conferences. There are some 35,000 journals, but they differ greatly in prestige, and nearly all important papers are published in about 3,500 of them. Conferences also vary. Most of the important discussion, however, takes place outside the main speeches. Owing to delays involved in formal channels, scientists, especially in invisible colleges, pass along more information by word of mouth.

The scientific community maintains research standards in various ways: education in a tradition, provision of research facilities, exercise of authority (usually informal), advice from colleagues, appraisal of papers before publication, and (especially) operation of the reward mechanism.

Because much scientific knowledge is highly formalized, an education in science tends to produce more intellectual conformity than most other types. As a graduate student, the scientist is required to specialize, and specialization tends to narrow his thinking. Formal authority rarely is exercised in science; informal authority, more often. Standards are maintained largely by the journals that publish scientific work. Contributions are judged by referees and sometimes by editors. Normally a scientist can get his paper published somewhere (assuming it is of minimal quality).

The reward mechanism of science is the granting of professional recognition in return for work that is original and meets accepted standards. It reconciles the self-interest of the individual scientist and the interest of the scientific community in advancing knowledge. Scientific rewards range from the informal (such as the frequent citation of a scientist's papers by other scientists) to the formal (such as the Nobel Prize). The reward mechanism is controlled mostly by recognized scientists. It operates fairly and rationally within limits—limits set by the fact that the most powerful judges (editors and referees) are the established scientists. Hence original, revolutionary work may be resisted or ignored. To prevent this, alternative metaphysical blueprints should be discussed ahead of time, so that radically new ideas are treated with respect when they eventually are submitted for publication.

The reward mechanism also speeds up research by intensifying competition among scientists. Scientists compete because they wish to be recognized as the first to make a discovery. However, competition not only increases anxiety, it also duplicates resources as several scientists vie with one another to solve the same problem. In previous cultures scientists did research mostly from love of learning. However admirable this might be, it would be hard to institute in a culture and society as geared to competition as ours. How diverse and how pervasive is the influence of the sociocultural world on science we shall now see.

NOTES

1. The term "invisible college" originally was applied in the seventeenth century to the informal predecessor of the Royal Society. It was revived by Derek J. de Solla Price in his *Science Since Babylon* (New Haven: Yale University Press, 1961), p. 99. On invisible colleges, see Belver D. Griffith and Nicholas C. Mullins, "Coherent Social Groups in Scientific Change," *Science* 177 (September 15, 1972): 959–64; Nicholas C. Mullins, "The Development of a Scientific Specialty: The Phage Group and the Origins of Molecular Biology," *Minerva* 10 (January 1972): 51–82, and *Theories and Theory Groups in Contemporary American Sociology* (New York: Harper & Row, 1973); and Diana Crane, *Invisible Colleges*.

2. From "bacteriophage," a virus which infects bacteria.

3. Mark Oliphant, *Rutherford: Recollections of the Cambridge Days* (Amsterdam: Elsevier, 1972), p. 108.

4. P. L. Kapitza, "Recollections of Lord Rutherford," in *The Physicist's Conception of Nature*, ed. Jagdish Mehra (Dordrecht, Holland, and Boston: Reidel, 1973), p. 757.

5. Oliphant, *Rutherford*, p. 29.

6. Kapitza, "Recollections of Lord Rutherford," p. 757.

7. Reprinted in *Phage and the Origins of Molecular Biology*, ed. J. Cairns, G. S. Stent, and J. D. Watson (Cold Spring Harbor, N.Y.: Cold Spring Harbor Laboratory of Quantitative Biology, 1966), pp. 9–22.

8. A. S. Eve, *Rutherford* (New York: Macmillan, 1939), p. 304.

9. Oliphant, *Rutherford*, p. 28.

10. J. G. Crowther, *British Scientists of the Twentieth Century* (London: Routledge & Kegan Paul, 1952), p. 71.

11. Kapitza, "Recollections of Lord Rutherford," p. 756.

12. See Mullins, "Development of a Scientific Specialty," pp. 61, 64. The invisible college recruits most of its new members from its own graduate students. Delbrück was a particularly able recruiter. Most academic scientists teach the graduate students who apply to them, and these in turn teach others, so that each new generation of students is only slightly larger than the generation that teaches it. This method keeps progress fairly slow for nearly 15 years. Delbrück cut short the waiting period by giving a summer phage course at Cold Spring Harbor aimed at scientists who already had students. By 1950 there were 35 people doing phage work, compared with four in 1945. With normal student recruitment this volume of research would not have been attained until 1960.

13. Cairns, Stent, and Watson, eds., *Phage and Origins of Molecular Biology*, p. 77. Cited by Crane, pp. 25–6.

14. Kapitza, "Recollections of Lord Rutherford," pp. 750, 758.

15. Mullins, "Development of a Scientific Specialty," pp. 72–73.

16. E.g., "Up the Crick with Watson," a five-verse doggerel recited at the 1953 Cold Spring Harbor Symposium. Robert Olby, *The Path to the Double Helix* (Seattle: University of Washington, 1974), p. 445.

17. Quoted by Mullins, "Development of a Scientific Specialty," p. 73.

18. Jerry Gaston, "Communication and the Reward System of Science: A Study of a National 'Invisible College,' " *The Sociological Review Monograph* 18, ed. Paul Halmos (University of Keele, 1972), pp. 27–8. Gaston quotes a British high-energy physics experimentalist as follows: "Publications are always too late. By the time a thing is published you ought to have known about it for six months. To be honest, I never read journals for this reason. I assume that anything which is worth my knowing has already come to my attention—big-headed—but that's life."

19. J. Martyn and A. Gilchrist, *An Evaluation of British Scientific Journals* (London: Aslib, 1968), p. 6.

20. Stephen Cole and Jonathan R. Cole, "Visibility and the Structural Basis of Awareness of Scientific Research," *American Sociological Review* 33 (1968): 397–413.

21. See Jagdish Mehra, *The Solvay Conferences on Physics: Aspects of the Development of Physics Since 1911* (Dordrecht, Holland, and Boston: Reidel, 1975).

22. F. W. F. Smith, *The Professor and the Prime Minister* (Boston: Houghton Mifflin, 1962), p. 43. Cited by Mehra, *Solvay Conferences on Physics*, p. xxii.

23. Quoted by R. W. Clark, *Einstein, the Life and Times* (New York: Avon, 1972), p. 185. Cited by Mehra, *Solvay Conferences on Physics*, p. xxii.

24. Quoted by Mehra, *Solvay Conferences on Physics*, p. xxii.

25. Quoted in *ibid.*, p. xiv.

26. *Ibid.*, p. xxvii.

27. Griffith and Mullins, "Coherent Social Groups in Scientific Change," 960, and Nicholas C. Mullins, "The Distribution of Social and Cultural Properties in Informal Communications Networks Among Biological Scientists," *American Sociological Review* 33 (1968): 786–97.

28. On science education as an apprenticeship in which the future scientist acquires knowledge and techniques, informal as well as formal, see Jerome R. Ravetz, *Scientific Knowledge and Its Social Problems*, ch. 3; Michael Polanyi, "The Tacit Component," in *Personal Knowledge*, pp. 69–245; and J. M. Ziman, *Public Knowledge: The Social Dimension of Science* (Cambridge: Cambridge University Press, 1948), ch. 4.

29. Thomas S. Kuhn, "The Essential Tension: Tradition and Innovation in Scientific Research," in *Scientific Creativity*, ed. Calvin W. Taylor and Frank Barron (New York: Wiley, 1963).

30. Liam Hudson, *Contrary Imaginations* (London: Methuen, 1966), and *Frames of Mind* (London: Methuen, 1968).

31. James D. Watson, *The Double Helix*, p. 86. Griffith was John (not Fred) Griffith, a young Cambridge mathematics graduate.

32. See Eugene Frankel, "Corpuscular Optics and the Wave Theory of Light: The Science and Politics of a Revolution in Paris," *Social Studies of Science* 6 (1976): 156–59.

33. Maurice Crosland, *The Society of Arcueil* (London: Heinemann, 1967), p. 92. Quoted by Frankel, "Corpuscular Optics and the Wave Theory of Light," 159.

34. Gerald Holton, "Striking Gold in Italy: Fermi's Group and the Recapture of Italy's Place in Physics," *Minerva* 12 (April 1974): 177–81; and Emilio Segrè, *Enrico Fermi: Physicist* (Chicago: University of Chicago Press, 1970), chs. 2, 3.

35. Harriet Zuckerman and Robert K. Merton, "Patterns of Evaluation in Science: Institutionalization, Structure, and Functions of the Referee System," *Minerva* 9 (1971): 66–100.

36. M. A. Libbey and G. Zaltman, *The Role and Distribution of Written Informal Communication in Theoretical High Energy Physics* (New York: American Institute of Physics, 1967), cited by Zuckerman and Merton, "Patterns of Evaluation in Science."

37. Ernst Mach, *History and Root of the Principle of Conservation of Energy*, trans. P. E. B. Jourdain (Chicago, 1911), quoted by Lewis S. Feuer, *Einstein and the Generations of Science* (New York: Basic Books, 1974), p. 273.

213

38. "The Matthew Effect in Science," in *The Sociology of Science: Theoretical and Empirical Investigations*, ed. Norman W. Storer (University of Chicago Press, 1973), pp. 439–59. See also Harriet Zuckerman, *Scientific Elite: Nobel Laureates in the United States* (New York: Free Press, 1977).

39. Jonathan R. Cole and Stephen Cole, *Social Stratification in Science* (Chicago: University of Chicago Press, 1973), ch. 7.

40. See M. Ross Quillian, "Physics as a Social Model," *Science* 183 (1974): 842–43.

41. See *ibid.*, 842.

42. Arthur Koestler, *The Sleepwalkers* (Harmondsworth, Middlesex: Penguin, 1964), p. 382.

43. Robert K. Merton, "Priorities in Scientific Discovery," *Sociology of Science*, p. 289: "These controversies, far from being a rare exception in science, have long been frequent, harsh, and ugly. They have practically become an integral part of the social relations between scientists."

44. Watson, *Double Helix*, pp. 102–4.

45. See F. Reif, "The Competitive World of the Pure Scientist," *Science* 134 (1961): 1957–62.

214

Chapter 9

The Sociocultural Background, Part I

Throughout its history, science has been affected less than most pursuits by forces in the social and cultural world in which it is done. Since the eighteenth century, especially, work in a mature science has called for specialized training in a sophisticated body of knowledge and its accompanying techniques, and has been directed to problems arising within the science itself. As a result, modern scientists form an intellectual elite claiming sole competence to judge one another's work.

Nevertheless, even if science rarely is determined, it often is influenced, by many factors acting in the culture and society of its time, on which in turn it exerts some influence of its own.[1] These factors range from world views and ideologies, through external patronage and educational systems, to economic forces and technology. In this and the next two chapters I will try to show how intimately these factors affect the growth of science, and how in turn science affects them.

WORLD VIEWS AND IDEOLOGIES

Science always is done in an intellectual environment that includes the world views and ideologies of a society. A world view is a set of beliefs, not open to empirical refutation, about the fundamental features of nature, or man-and-society, or both. An ideology is a set of beliefs about how men should live and act in a world of this kind.[2] Because they permeate the thought and feeling of societies and

215

classes, world views and ideologies have considerable influence on the life of science.

A world view and its accompanying ideology may be held by an entire society or by a particular social group. Few individuals subscribe to a world view and ideology in their entirety, but each individual does so enough to feel united with others who do likewise. All cultures and societies are pervaded by polarities and antagonisms, such as those between freedom and order, youth and age, the family and the state. By means of a world view and its ideology most strains in the social and cultural order may be explained, justified, or criticized. With the aid of a world view and ideology, the individual can make sense of his world, endure its stresses, and act in it with purpose.

In times of social and political crisis, the traditional world view and certain rules of life are called into question by the group or class which precipitated the crisis, and alternatives are sought to reinforce or replace them. When traditional assumptions no longer explain the disruptions in their lives, men become perplexed and confused—as does the scientist in both his private life and his work. He is driven to examine his fundamental assumptions, test the scientific assumptions on which they have a bearing, and see his work in some relation to the crisis of his time. Thus social and political crises, involving world view conflicts, often contribute to crises in science and assist the rise of powerful new research traditions.

By way of example let us look at the interaction between Darwin's theory of evolution and the conflicting conceptions of man's place in the world held in nineteenth-century Britain.[3] During this period Britain was transformed from an agrarian, aristocratic, relatively harmonious society into one that was urban, industrial, and strife-ridden. The world view and ideology by which the aristocracy explained and justified the earlier social order were undermined by the world view and ideology through which the newly powerful, commercial middle class explained and justified the social conditions it had helped create. The earlier social order had been justified by the view that in nature and society alike there is a divinely ordained harmony: in nature, the hierarchy of fixed species forming the Great Chain of Being with man at its apex; in society, the divinely sanc-

216

tioned hierarchy of social classes. But the Industrial Revolution, with its social dislocations and its mass migrations of labor, discredited this scheme. The new social conditions then came to be justified by a new view of nature and society. Nature was seen as an arena in which species struggle for scarce resources in accordance with the iron laws of natural selection and the survival of the fittest. Society was regarded as a place in which individuals compete for jobs, and firms compete for markets, according to the rigid laws of Malthus and Adam Smith. Social inequality, social struggle, middle-class power, and laissez-faire capitalism all were said to reflect the inexorable laws of nature itself. By such thinking the industrial middle class justified its opposition to state intervention in the economy.

The conflict of world views was particularly evident in the debate between science and religion. Most theologians and scientists defended the existing social order, but where theologians saw social inequalities as God-given and sanctioned by religion, scientists regarded them as biological in origin and explained by science. Neither side thought of social inequality as a political problem calling for a political solution.

Darwin's theory played an important part in the transition from one world view to the other. Where Darwin had interpreted nature by analogy with society as explained by laissez-faire economics, later social theorists interpreted society by analogy with nature as explained by Darwin. As I have noted, Darwin was influenced by the emerging world view and ideology, in the sense that these made Malthus and Adam Smith important interpreters of social life and so key sources on which to draw for analogies with nature.[4] Darwin then contributed decisively to the development of this world view and ideology by proposing a scientific theory of nature that used economic precepts to account for a vast array of biological facts and therefore seemed to make the social scene reflect inevitably the state of nature. Thus the notion of nature, explained with the aid of economic ideas, was used to explain and justify the social order which had been explained and justified by these ideas in the first place. Laissez-faire economics and Darwinian evolution reinforced each other in the fully formed world view and ideology of late Victorian capitalism.

It has been argued that although economic ideas were impor-

217

tant to Darwin in forming his theory, they did not persuade his scientific colleagues to accept it. Instead, his peers were moved by the theory's success in organizing and explaining a multitude of troublesome facts in a range of fields such as geology, paleontology, zoogeography, and plant and animal classification.[5] However, this argument overlooks the fact that the emerging world view and ideology were powerful enough to lead Darwin to draw on nonscientific ideas, to which many of his colleagues were unfriendly, and to plant these ideas in the heart of a theory that finally persuaded them. In accepting the theory, Darwin's colleagues accepted his use of the ideas, if not the ideas themselves. This fact alone shows the influence of a world view and ideology on the science of the time.

As an example of a scientist influenced in his work by ideological considerations, consider Alfred Russel Wallace, who arrived at the idea of evolution by natural selection independently. Wallace was a socialist. In the 1860s he realized that in Darwin's theory natural selection acted by chance alone and was in no sense "progressive." He found this notion incompatible with his belief that man is inherently destined to build a just society. So he rejected natural selection as a sufficient explanation of the origin of men. Darwin then called him a "renegade to natural history."

PHILOSOPHY

Unlike world views, philosophies are rigorously formulated systems of ideas. Many assumptions by which scientists are guided, such as notions of space, time, matter, and causality, already have been analyzed by philosophers. Hence, scientists often have found both inspiration and clarification in philosophy.

For example, the most powerful driving force in German science during the first half of the nineteenth century was idealist philosophy, especially its theory of nature (*Naturphilosophie*). Many scientists accepted Hegel's and Schelling's theory that nature is a manifestation of an evolving Absolute or World Spirit. Thus Friedrich Kielmeyer, a biologist, proposed the law of embryological

218

recapitulation. Since, he argued, higher organisms possess certain characteristics of lower ones, and since all organisms come from the same source, every organism develops as an embryo through stages in which it resembles organisms below it in the scale of beings. Cell theory—also the creation of German scientists—was guided by the thesis of the philosopher Lorenz Oken that all living things are composed of tiny vital units. Scientists identified these units with the cells that were observed through the new achromatic microscopes in the 1830s. In his long search for the connection between magnetism and electricity (ended successfully in 1820), Hans Christian Oersted, a disciple of Schelling, was sustained by the view that both forces are aspects of World Spirit, and that in the phenomenon of polarity they exemplify the dialectical tension between apparent opposites through which World Spirit evolves. In the 1840s, in thermodynamics, Robert Mayer and Hermann von Helmholtz were moved by the idealist notion that matter ultimately is force and that all force is conserved.

In our own century Niels Bohr, who proposed that the electron "jumps" discontinuously between energy levels, was intrigued by Kierkegaard's theory that the individual "leaps" from one moral state to another; Heisenberg was guided by Plato's theory that the ultimate constituents of the universe are mathematical relations; and Einstein was impressed by Ernst Mach's analysis of the Newtonian concepts of space and time. This certainly is not to say that all scientists have read philosophy with interest. To some, like Rutherford, Dirac, Fermi, and Crick, formal philosophy meant little or nothing.[6]

However, nearly all scientists are moved by principles of a metaphysical kind, though they may not always recognize them as such. These principles are empirically untestable hypotheses about the ultimate constituents of the world. They suggest to scientists that certain entities and interactions exist and that theories may be developed to describe them and to predict their observable effects. These principles are limited in number and most of them recur in the history of science. They include *reductionism* (that all phenomena ultimately are caused by the behavior of a single class of entities, normally those of physics) and *emergentism* (that the universe consists of different levels of organization, and that the properties of higher or

219

more complex levels cannot be explained in terms of the properties of lower or less complex ones). In biology we meet the principles of *mechanism* (that all organic phenomena ultimately are explicable by means of physicochemical laws) and *vitalism* (that living things contain a vital element that cannot be explained as a property of inanimate matter). In physics we find such principles as *atomism* and *continuism*.

The influence of these last two principles can be clearly traced in the history of science. *Atomism* is the thesis that the world consists ultimately of discrete, relatively indestructible units, occupying points or tiny regions of space, and that all change can be explained in terms of the rearrangement of these particles in space and time. *Continuism* is the thesis that matter is present everywhere in the universe as a single, continuous element—the plenum—of which all particular bodies are manifestations. Atomism was proposed in antiquity by the pre-Socratic philosopher, Democritus, and was elaborated by the Epicureans. Continuism was founded by Parmenides and worked out systematically by the Stoic school. One feature of continuism, that nature abhors a vacuum, was adopted by many Greek scientists, notably Aristotle. In the early modern period continuism was espoused by Descartes, Leibniz, and Huyghens (who proposed that light travels like a wave through a continuous medium). Atomism was advocated by Newton. In fact, Newtonian atomism dominated physics until the mid-nineteenth century. Then continuism returned in Faraday's theory of the electromagnetic field, which was mathematized by Maxwell and united by him with the wave theory of light (revived by Young and Fresnel earlier in the century). Continuism then was adopted by Einstein in his special and general theories of relativity. In quantum mechanics, on the other hand, atomism and continuism have been combined unsatisfactorily, subatomic entities being explained as both particle and wave phenomena but not as both simultaneously. The greatest task facing physicists today is to reconcile the atomist and continuist approaches, or to replace one with the other, so as to produce a coherent quantum theory of microphenomena consistent with the relativistic theory of macrophenomena.

RELIGION

Like science, religion is a sociocultural movement as well as a source of ideas. Religions have influenced science in two ways: by furnishing individual scientists with profound beliefs about man and the world, and by providing organized support for, or opposition to, scientific research.

Although the early Church tended to be otherworldly, rejecting Greek paganism and Roman hedonism, the Christian religion on the whole has been well disposed to scientific inquiry. The belief that the world was created for man's sake by a rational God authorizes man to look in nature for God's handiwork and explain it with theories that are (in Copernicus's words) "pleasing to the mind." In the Middle Ages, having adopted a rational theology based on Aristotle, the Church encouraged scientific inquiry on Aristotelian lines. In the early Renaissance such great scientists as Copernicus and Galileo were Catholics and Kepler was a Lutheran. Galileo was inspired, in fact, by the religious belief that God had written the book of nature in the language of mathematics. It was not until he challenged the Church's authority to decide between the Ptolemaic and Copernican theories that the Church "secluded" him and actively discouraged similar scientific research.

By insisting on the literal truth of the Copernican theory, Galileo attacked the fundamental assumption of the Aristotelian cosmology, with which the Church supported its theology—the assumption that there is an absolute difference between the perfect heavens and an imperfect earth. For if Copernicus were right, and the earth moved round the sun with the other planets, one no longer could claim that the planets were perfect and the earth not. Galileo took his case to the public in a series of brilliant, polemical books and pamphlets, in one of which he had the temerity to put the Pope's own scientific ideas into the mouth of a simpleton. When Galileo insisted on publicly refuting the theologians in a domain they regarded as their own—cosmology (at that time, natural theology)—the theologians turned against him.

Thereafter, leadership in science passed to Protestant-

221

dominated Northern Europe. In seventeeth-century England the Puritans strongly supported scientific inquiry. They believed that the millennium was coming and that man must prepare for it by regaining the physical and intellectual powers he had lost at the Fall. With the aid of science and technology he would make himself a new Adam and the world a second Eden.[7] Most members of the Royal Society were deeply religious and believed that science, like theology, was a way to prove the existence and beneficence of God. Newton, for instance, sought to show that God was "active" in the world. He argued that the universe and its constituent bodies consisted mainly of empty space across which gravity and other forces acted instantaneously. He contended that, in the absence of a material medium to carry them, the effects of these forces had to be transmitted through the agency of God Himself.

During the eighteenth century, French scientists and mathematicians, less religiously inclined, ignored Newton's theological motive and dismissed the question of what caused these forces to operate. Instead they sought to express, in precise mathematical form, the laws by which the forces were governed. In this approach they were encouraged by Bernard de Fontenelle, perhaps the greatest of all popularizers of scientific ideas. Fontenelle was a sceptic. In presenting scientific ideas to the public he implied that the Church, in France especially, was the archenemy of disinterested science. His role was inherited by Voltaire, who dismissed the Christian religion as a delusion and asserted that the way to truth was through science. Thus, under the Encyclopedists, the world came to be seen as a machine running according to permanent laws with no need for divine intervention. When Napoleon asked Laplace whether God had created the planetary order described in his celestial mechanics, the scientist replied that he did not need that hypothesis. With the Revolution, science was institutionalized in France as a secular enterprise.

In England, on the other hand, science continued to be regarded as the discovery of God's design in nature, though that design was variously interpreted. Certain chemists, for example, particularly from the dissenting sects, rejected Newton's theory of matter. Among them was John Dalton, who believed that God would have created a solid, substantial universe rather than one filled with empty space. Dalton argued against the Newtonian view that the chemical elements (hydrogen, sulfur, mercury, and so on) are composed of homogeneous particles in different arrangements with void space between them, and proposed instead that the atoms of each element are solid, distinct, and indivisible.[8]

The last great challenge to religion came with the publication of Darwin's theory of evolution. As in the case of Galileo, the theologians claimed that Darwin had trespassed on their domain and contradicted the Bible. The world, they said, had been created much more recently than Darwin allowed, and man had been made the lord of nature instead of emerging by chance from monkeys. However, the scientific evidence for Darwin was overwhelming, and by the end of the century his theory had been accepted by many religious scholars, who regarded the evolution of species as an ascent toward man designed by God himself. (They either ignored, or disagreed with, Darwin's contention that all species appear by chance and that evolution has no direction.)

Today the theory of evolution is accepted by most religious scientists, and religion and science generally are regarded as complementary rather than conflicting interpretations of nature. Science, on this view, investigates the physical world, while religion gives meaning to man's life. Indeed, as I read history, scientific thinking consistently has had much in common with theology. Science's greatest theories have been cosmological—that is, concerned with the ultimate nature of the universe as a whole—and some of the greatest scientists of recent times (witness Faraday, Maxwell, Planck, and Einstein) have been in the broadest sense religious men.

NATIONAL SCIENTIFIC TRADITIONS

Every economically developed nation has a scientific tradition of its own. In Western Europe, national scientific traditions were particularly marked during the eighteenth and nineteenth centuries before the arrival of a more uniform, transnational culture. In England, where the middle class could pursue its commercial interests in relative freedom, the investigation of nature tended to be empirical and experimental. It was concentrated in the sciences that had risen from the medieval crafts and technologies—electricity, heat, magnetism, optics, and chemistry. Influenced by Bacon's dictum that "knowledge is power," the Royal Society dealt mainly with practical problems and applied physics. (Newton was not a typical member.) In the nineteenth century, much science was done under the auspices of specialist societies with a practical bent, and around new industrial centers like Manchester and Birmingham. The societies brought science to the masses through popular lectures. Science, it was thought, should be intimately connected with the needs of everyday life.

In eighteenth-century France, under a feudal regime, the middle class was more restricted. Scientific research was directed largely by the government through the Académie des Sciences, a more aristocratic institution than the Royal Society and more devoted to the interests of its kingly sponsors. French education, too, was strongly influenced by the philosopher Descartes's insistence on clarity of thought and expression. Science, then, tended to be more abstract than across the Channel and less concerned with phenomena met in practical pursuits. Where English scientists were influenced mainly by Newton's *Optics* (his more empirical work), a succession of French mathematicians and physicists recast the concepts and mathematics of Newtonian mechanics. In contrast to the empirical studies of British chemists, Antoine Lavoisier produced a general theory of chemistry. After the Revolution, French science retained its mathematical cast and became concentrated in a few great institutions, such as the École Polytechnique. Whereas an Englishman, James Watt, had built the steam engine, the French engineer Sadi Carnot wrote a theory of the ideal heat engine.[9]

In Germany, feudal and monarchical regimes lasted much longer, and supported a philosophy of national idealism. In the universities, science was taught and practiced alongside the great philosophic systems of Kant, Fichte, Schelling, Hegel, and Lotze. Hence, until the mid-nineteenth century at least, scientists tended to propose large-scale theories drawing not only on a range of sciences but also on philosophies. Leonhard Euler and Immanuel Kant proposed holistic field theories, Clausius and Helmholtz developed the conservation laws of thermodynamics (uniting all forms of energy), Matthias Schleiden and Theodore Schwann founded cell theory (treating the cell as an independent whole), and biologists subscribed to vitalism (the notion that an organism is a whole unified by a "vital spirit").

In the nineteenth century, French and German scientists often were bitter rivals. Justus Liebig, for instance, criticized the claims of French chemists. By introducing a new nomenclature, Lavoisier and his colleagues had "blotted out" the achievements of the phlogistic chemists Joachim Becher and Georg Stahl, so that "To many the knowledge we now possess appears to be only the inheritance of the French school of that day."[10] Liebig called attention to German contributions to chemistry, such as Jeremiah Richter's law of equivalent proportions,[11] saying, "The discovery of that law must be attributed to the sagacity and acuteness of a German chemist and the name Richter will remain as imperishable as the science itself."[12]

Even today national differences persist in science, though less strongly. An empiricist tendency still prevails in British science. Writing in 1906, the French philosopher-physicist Pierre Duhem remarked that British physicists liked to equip their theories with mechanical models.[13] Ernest Rutherford at Cambridge gave British physics a decidedly experimental cast. Where Germany had three great quantum theorists—Planck, Heisenberg, and Schrödinger—Britain had one, Dirac. In the United States, science was (and still is) markedly experimental and practical. Albert Michelson and Robert Millikan[14] were admired while the theorist Willard Gibbs was ignored.[15] The American genius for applied science was in keeping with American commercial inventiveness and zest for expansion. Today, in the Soviet Union, research being done in Newtonian mechanics reveals the impact of the official philosophy of materialism. Similarly, the development of cybernetics (resisted at first as counterrevolutionary) into a synthesizing discipline in science and technology reflects the interests of a totalitarian government in economic planning and political control. Nowhere has science been made to serve a nation more than in the Soviet Union (see chapter 10).

PUBLIC OPINION

Science always has been sensitive to public opinion. The more science is esteemed, the more young people choose it as a career, flocking to those sciences that are most admired. When the public becomes disillusioned, recruits fall off, and scientists begin to question the value of their research. Above all, public opinion is reflected in the amount of money science receives. Here are some examples of the influence of public opinion on science.

Late in the eighteenth century, after the Terror, the leaders of the French Revolution hailed science as a rational source of true knowledge and useful invention. Science education was expanded, and top scientists were hired as teachers. Napoleon encouraged scientists in many ways, showing them personal favor, offering prizes for work in such areas as voltaic electricity, and giving a huge electric pile to the École Polytechnique. As a result French science led the world for three decades. With the restoration of the Bourbons in 1814 the public mood changed and a new cultural style took over. Not science and mathematics but literature, politics, and oratory were now the fashion.[16] Scientists in turn sought to please the public by lecturing in declamatory style. Many scientists went into academic or government administration, and some (such as Arago and Gay-Lussac) tried politics. Between 1815 and 1870 the government financed only two projects in physics, both of them investigations of the properties of steam and gas, done to increase the safety and efficiency of steam engines. Throughout this time German science went from strength to strength.

Now consider the United States where, during the nineteenth century, scientists were thought to be no more intelligent than the next man. Americans regarded science mainly as a form of practical experimentation, and pure science languished. An attempt was made to replace Willard Gibbs at Yale (where he was unpaid) on the grounds that his theoretical work in thermodynamics and statistical mechanics was of no practical use. Inventors, on the other hand, such as Samuel Morse, Thomas Edison, and Alexander Graham Bell, were widely admired. Not until the beginning of the twentieth century was science seen as a profession producing knowledge of

226

social importance. After World War II, won with the aid of basic research, science came to be seen as a unique source of technologies to raise living standards and supply arms for the cold war. The public encouraged governments to spend heavily on science, especially nuclear and particle physics, and more and more young people became scientists. Late in the 1960s, however, the popular identification of science with physics, and of physics with warfare and technology, led to a mood of disenchantment. Students began to prefer the life sciences to physics and chemistry. Scientists themselves went through a crisis of conscience and sought to make their studies more socially responsible by speaking out on the misuses of science in war and industry.

A striking example of the influence of public opinion was the abandonment of strict causality by physicists in Weimar Germany in response to the hostility of intellectuals to mechanistic science.[17] This example also illustrates the influence on science of world-view conflicts and political factors, e.g., military defeat. In Germany after World War I three world views were in conflict: dialectical materialism, deterministic materialism, and *Lebensphilosophie* ("philosophy of life"). Marxian dialectical materialism had little support, but deterministic materialism, the world view of the business community and most physical scientists, was widely held. It maintained that the physical world is governed entirely by mechanistic laws (laws that exclude goals and purposes), which can be discovered by science and exploited by technology to improve the human condition immeasurably. (Deterministic materialism arose in the 1840s as a reaction of the upper classes to the Romanticism, and in science *Naturphilosophie*, by which, earlier in the century, those same classes had sought to explain and justify the comparative social and economic backwardness of their country. Deterministic materialism both contributed to, and later was reinforced by, the unification of the German states and the scientific-technological alliance established by German industry in the second half of the century.)

By the time of the Weimar Republic, however, deterministic materialism had lost its appeal, since the industrialists, together with their world view and ideology, had been discredited by Germany's military defeat. Intellectual and academic circles turned eagerly,

227

then, to a world view and ideology that valued feeling, spontaneity, and community over logical analysis, quantification, and bureaucracy. This world view, in essence a revival of Romanticism, was expressed in *Lebensphilosophie*, neovitalist biology, and Gestalt psychology. It was reinforced by Oswald Spengler's enormously influential work, *The Decline of the West*, which declared that Western mechanistic science was exhausted and that a new quasi-religious, emotional sensibility was inevitable. Intellectuals and academics welcomed Spengler's view that science and learning in Germany had reached a crisis, and they looked for a vitalist revolution to sweep through all branches of knowledge.

German physicists, too, had read Spengler. Under his influence, those opposing mechanism declared that their science was in a crisis, which they actually welcomed. They also were disturbed at losing the public esteem they once enjoyed. Instead of retiring into their science, however, they sought to make it respond to the criticisms of other intellectual communities. This is not to say that they abandoned their scientific convictions to placate public opinion, but rather that many of them sympathized with this opinion and hence changed their science more radically than they would have done without it.

Quite suddenly, then, in the aftermath of World War I physicists began to abandon strict determinism and to announce that the laws of science ultimately were statistical, thus making room for chance in inorganic nature. In 1919 Franz Exner and Hermann Weyl came out against causality; in 1921 they were joined by Walter Schottky, Richard von Mises, and Walther Nernst; in 1922, by Erwin Schrödinger (who returned to causality, however, in his wave mechanics of 1925); and in 1925, by Hans Reichenbach. Between 1922 and 1925, largely for internal reasons, the "old quantum theory" of the Bohr-Summerfeld research tradition was plagued by serious difficulties. As a way out, the Bohr-Kramers-Slater theory, proposed in Copenhagen (1924), explained the interaction of matter and radiation statistically and acausally. The theory was welcomed at once by German physicists opposed to causality and mechanism.

This retreat from causality took place before 1926, the year in which Heisenberg proposed his nondeterminist matrix mechanics and Max Born interpreted Schrödinger's wave mechanics statistically—both events which would have justified the retreat scientifically. It is noteworthy that except for Einstein, the leading physicists who upheld causality—Planck, Schrödinger, von Laue—were either political conservatives or interested in classical literature. Those physicists who opposed causality—Nernst, von Mises, Born, Weyl—were either political liberals or interested in contemporary literature. Thus the widespread abandonment of causality in German physics during the Weimar period seems to have been the result of a deliberate decision by many physicists to adapt their science to external criticisms with which they were in sympathy.

EDUCATION

The growth of science may be greatly accelerated or retarded by a country's educational system. In the nineteenth century, as I have said, both French and German science leaped forward as a result of educational reforms carried out with scientific growth in view. In France the Revolution began by abolishing the scientific institutions of the *ancien régime* with their tradition of abstract mathematical science. But after the fall of Robespierre the rationalism to which the Revolution was committed prevailed. The École Polytechnique and the École Supérieure were created, and many earlier institutions were revived. This work was sustained by a Baconian belief in the power of science to improve the moral and material condition of mankind. Leading scientists, such as Laplace and Lagrange, taught in the Écoles, and their students regarded themselves as the intellectual vanguard of civilization. Under Napoleon, however, science came to be valued mainly for its utilitarian possibilities, especially in engineering and warfare. French science graduates were mathematically among the most sophisticated in the world, but many saw their science mainly as a source of practical knowledge. This was one cause of the decline of pure science in France after its burst of creativity early in the century.

In Germany, beginning in the 1820s, a state system of scientific education was created that made German science preeminent

229

from the 1830s to World War I. Over 20 universities, all subsidized by the state, together turned out five times as many graduates as the universities of Britain. Most of them possessed thriving schools of scientific research, which attracted students from all over the world. The excitement of German science in those days is conveyed by Justus von Liebig, who in 1826 at the University of Giessen was the first to offer systematic training in chemical research:

> A kindly fate had brought together in Giessen the most talented youths from all countries of Europe. . . . Every one was obliged to find his own way. . . . We worked from dawn till night; there were no recreations and pleasures at Giessen. The only complaints were those of the attendant, who in the evenings, when he had to clean, could not get the workers to leave the laboratories.[18]

Early in the nineteenth century, German education was dominated by the humanities. In order to justify themselves, scientists claimed that they too sought knowledge for its own sake and trained the intellect as an end in itself. Instead of being separated from other studies, as in the École Polytechnique, science was taught in the Philosophische Facultät (philosophy department). German science, then, came to maturity in a philosophic milieu which encouraged rigorous and comprehensive thinking and stimulated the creation of scientific theories. German universities also encouraged scientific research by rewarding productive professors with rapid promotion and higher salaries. As the century wore on, the universities threw themselves single-mindedly into research and the training of researchers. The ideal held up to student and professor alike was a selfless devotion to the advancement of knowledge. This ideal is typified in the "Academical Discourse" given by Hermann von Helmholtz at Heidelberg in 1862. "In conclusion" declared Helmholtz,

> I would say, let each of us think of himself, not as a man seeking to gratify his own thirst for knowledge, or to promote his own private advantage, or to shine by his own abilities, but rather as a fellow-laborer in one great common work bearing upon the highest interest of humanity.[19]

230

It is harder to generalize about the United States, where there is no central ministry of education and where local control and private initiative prevail. Nevertheless, it is undeniable that until the end of the nineteenth century American education neither reflected nor contributed significantly to the growth of pure science. It was not until the 1890s, for example, that major universities awarded Ph.D.'s in physics and chemistry. In fact, science education expanded mainly in response to external demands—from government, industry, agriculture, and the military—for information and techniques necessary for national survival and technological growth.

Indeed, it was only when Americans were suddenly confronted with Russian space achievements in the 1950s that support of basic science education was made a national priority. An alarmed Congress hurriedly created the National Science Foundation, giving it power to grant funds to any scientific organization with a project that the foundation thought likely to contribute to basic science and national prestige. At the same time, in almost every school system, science education was radically and hastily reformed. Designed by scientists, the early reforms stressed cognitive learning. Today, however, science instruction is planned mostly by teachers. It has become less specialized and considers the place of science in daily life.

In my view, science education in the United States at the prespecialization level still is too abstract and too academic; it appeals to the budding scientist rather than to students in general. Too many teachers fail to win the interest of young people who are not scientifically inclined but would like to study natural phenomena. At both secondary and higher levels, teachers generally present science as a fixed body of knowledge and as a method of inquiry which they demonstrate under restricted conditions. To arouse general interest in science, they should present it whole—as a historical movement, a community of scholars, and a cultural force, as well as a method of inquiry and body of knowledge. They should pay far more attention to how, by whom, and with what motives scientific discoveries were made. They should stress the increasing importance of science in the history of ideas and in the evolution of the characteristic modes of thought and feeling of particular civilizations.

231

If teachers at all levels were to describe the activities of the scientific community, they would show more convincingly that science is a cooperative and very human endeavor that seeks to attain a progressively more accurate and comprehensive understanding of the universe by submitting its knowledge to continual scrutiny. If they were to present science as a cultural force, they would make their students more aware of the place of science in human life as a whole.[20]

I am, of course, advocating that teachers adopt the view of science expounded in this book. I believe that it would lead to a more humane and stimulating science education.[21]

SUMMARY

Science is influenced by a range of factors in the culture and society in which it is done. These include world views and ideologies, philosophy, and religion.

New world views and ideologies, arising in social and political turmoil, often contribute to revolutions in science. Darwin, for example, reflecting on evolution, was influenced by the emerging world view and ideology of laissez-faire capitalism. His theory then reinforced that world view, giving it in addition a biological cast.

Scientists also have been guided by philosophy. Idealist *Naturphilosophie*, for instance, inspired most research done in Germany during the first half of the nineteenth century. Scientists also make metaphysical assumptions of their own, such as those undergirding atomism and continuism.

There always has been an interplay between science and religion. Christianity on the whole has favored science, but there have been serious conflicts between them, notably over Galileo's defense of Copernicanism and Darwin's theory of evolution.

In the sociocultural environment, I considered three factors: national scientific traditions, public opinion, and education.

National scientific traditions were especially strong in the eighteenth and nineteenth centuries, but they persist even in this day of international science. I mentioned in particular the British empirical, the French mathematical, and the German holistic, traditions.

Science and its patrons are highly sensitive to public opinion. French science bloomed under Napoleon but later went out of favor and into decline. In the United States pure science was largely ignored until after World War II. It then became the knowledge base of technology and grew dramatically as money and recruits poured in. In Germany after World War I many physicists abandoned strict causality in response to public revulsion against mechanistic science.

Education can do much to help or hinder science. During the nineteenth century, educational reforms in France and Germany, especially at the university level, made those countries the successive leaders of European science. In the United States, science education mainly served applied science. Only the evidence of Soviet space triumphs brought home to American leaders the importance of education in basic science. Although this deficiency in part has been overcome, science teachers still fail to portray science as a vital element of a total culture.

In the next chapter I will look at political and economic factors in the growth of science. I also will consider the impact of science on society and culture.

NOTES

1. *Society* may be regarded as the network of human interactions, and *culture* as the framework of beliefs, symbols, and values by which these interactions are interpreted. See Clifford Geertz, *The Interpretation of Cultures: Selected Essays* (New York: Basic Books, 1973), pp. 144–45: "One of the most useful ways . . . of distin-

guishing between culture and social system is to see the former as an ordered system of meaning and of symbols, in terms of which social interaction takes place; and to see the latter as the pattern of social interaction itself. On the one level there is the framework of beliefs, expressive symbols, and values in terms of which individuals define their world, express their feelings, and make their judgments; on the other level there is the ongoing process of interactive behavior, whose persistent form we call the social structure." In this chapter I focus on the intellectual component of culture.

2. A world view differs from a metaphysics in that it rarely is analyzed rigorously and may not even be explicitly formulated. An ideology differs similarly from an ethics.

3. See Robert Young, "The Historiographic and Ideological Contexts of the Nineteenth-Century Debate on Man's Place in Nature," in *Changing Perspectives in the History of Science*, ed. Mikuláš Teich and Robert Young, pp. 340–438.

4. As John Maynard Keynes put it, "the principle of the Survival of the Fittest could be regarded as one vast generalization of the Ricardian economics." *Collected Writings:* vol. 9, *Essays in Persuasion* (London: St. Martin's, 1972), p. 276.

5. Thomas S. Kuhn, "The Relations Between History and History of Science," *Daedalus* 100 (Spring 1971): 281–82.

6. The following conversation is said to have taken place between Rutherford and the philosopher Samuel Alexander. Rutherford: "When you come to think of it, Alexander, all that you have said and all that you have written during the last thirty years—what does it all amount to? Hot air! Hot air!" Alexander: "And now, Rutherford, I am quite sure that you will like me to tell you the truth about yourself. You are a savage—a noble savage, I admit—but still a savage!" A. S. Eve, *Rutherford* (New York: Macmillan, 1939), p. 240. One of my own colleagues, a renowned physicist, told me that the philosophy of science was no more than "yesterday's physics."

7. Charles Webster, *The Great Instauration: Science, Medicine and Reform 1626–1660* (London: Duckworth, 1975).

8. See Arnold Thackray, *Atoms and Powers*.

9. Thomas Kuhn contrasts British and French science in the 18th and 19th centuries in "The Relations Between History and History of Science," pp. 279–80, and "Scientific Growth: Reflections on Ben-David's 'Scientific Role,'" *Minerva* 10 (1972): 173–74. See also A. Rossi, "The Two Paths of Physics (First Part)," *Scientia* 108 (1973): 565–83. On the differences between British, French, and German science, see M. P. Crosland, "Introduction," in *The Emergence of Science in Western Europe*, ed. M. P. Crosland (London: Macmillan, 1975), pp. 1–13. On the differences between the Newtonian (British), Cartesian (French), and Leibnizian (German) research traditions, especially in the eighteenth century, see Yehuda Elkana, "Newtonianism in the Eighteenth Century" (essay review), *British Journal for the Philosophy of Science* 22 (1971): 297–306. On the scientific institutions influencing

the traditions I have described, see Joseph Ben-David, *The Scientist's Role in Society*, chs. 5–7, and Yehuda Elkana, *The Discovery of the Conservation of Energy* (Cambridge: Harvard University Press, 1974), ch. 6.

10. *Letters in Chemistry*, 3d ed. (London, 1851), p. 26. Quoted by Crosland, "Introduction," p. 4.

11. In 1791 Richter proposed that the weight of a substance A, combined with a given amount of a substance B, also would combine with that weight of a substance C which entered into combination with the same amount of substance B.

12. *Letters in Chemistry*, p. 96. Quoted by Crosland, "Introduction," p. 4. National rivalry often became personal, as in the case of Robert Koch and Louis Pasteur, both of whom advanced immunization. In my view, Koch's work was as important as Pasteur's, though it is less well known. True, Pasteur discovered how anthrax was transmitted (by worms that fed on the dead bodies of diseased sheep buried deep in the ground and then rose slowly to the surface). But Koch already had reached the same conclusion through his own laboratory experiments, and he accused Pasteur of plagiarism. Koch's group also did more than Pasteur and his assistants to isolate other microbes, as in the case of cholera, where the Frenchmen failed miserably (leading to the death of one of them) and the Germans succeeded. Koch was a better bacteriologist, it would seem, but Pasteur was more daring. The Frenchman risked a murder charge by being the first to inject a rabies vaccine (which he had invented) into a human being

13. *The Aim and Structure of Physical Theory*, trans. P. P. Wiener (New York: Atheneum, 1962), p. 72: "The French or German physicist conceives, in the space separating two conductors, abstract lines of force having no thickness or real existence; the British physicist materializes these lines and thickens them to the dimensions of a tube which he fills with vulcanized rubber."

14. In a series of experiments between 1913 and 1917, Millikan measured the motions of charged oil drops to find the charge of the electron. From the amount of this charge he deduced that an electron was 1,836 times lighter than a hydrogen ion, thus confirming the prediction of British physicist J. J. Thomson.

15. An incident recounted by J. J. Thomson reveals Gibbs's isolation: "When a new University was founded in 1887 the newly elected President came over to Europe to find Professors. He came to Cambridge and asked me if I could tell him of anyone who would make a good Professor of Molecular Physics. I said, 'You need not come to England for that; the best man you could get is an American, Willard Gibbs.' 'Oh,' he said, 'you mean Wolcott Gibbs,' mentioning a prominent American chemist. 'No, I don't,' I said, 'I mean Willard Gibbs,' and I told him something about Gibbs's work. He sat thinking for a minute or two and then said, 'I'd like you to give me another name. Willard Gibbs can't be a man of much personal magnetism or I should have heard of him.' " *Recollections and Reflections* (New York: Macmillan, 1937), pp. 185–86.

235

16. Robert Fox, "Scientific Enterprise and the Patronage of Research in France 1800–70," *Minerva* 11 (1973): 442–73.

17. Paul Forman, "Weimar Culture, Causality, and Quantum Theory, 1918–1927: Adaptation by German Physicists and Mathematicians to a Hostile Intellectual Environment," in *Historical Studies in the Physical Sciences*, ed. Russell McCormmach, vol. 3, pp. 1–115.

18. Quoted by Eric Ashby, *Technology and the Academics: An Essay on Universities and the Scientific Revolution* (New York: St. Martin's Press, 1966), p. 24.

19. Quoted in *ibid.*, p. 26.

20. For an excellent conceptual analysis of scientific knowledge and inquiry, of special value for science teachers, see James T. Tobinson, *The Nature of Science and Science Teaching* (New York: Wadsworth, 1968).

21. It would take too long to justify my view of science education here. Suffice it to say that in addition to being taught as tentative rather than final, scientific knowledge should be presented in terms of laws and theories rather than topics or problems. In physics, for example, such topics as mechanics, heat, light, sound, electricity, and magnetism, should be taught not in isolation but as interrelated by thermodynamics, electromagnetics, and relativity. Topic-centered courses do not permit long-term development of scientific ideas. Each topic generally is treated as a sovereign entity, not necessarily deriving from or leading to another, but it cannot be understood properly without a grasp of the basic knowledge on which it rests. Also, scientific method should be presented as a driving force of science. Instead of performing laboratory exercises imitating or confirming the teacher's demonstrations, students should do laboratory work that is genuinely experimental. If a student is to learn how to think and act scientifically, he must do some research of his own.

Chapter 10

The Sociocultural Background, Part II

In the English-speaking world, scientists generally have accommodated themselves to governments in power. As early as 1663, in his draft preamble to the statutes of the Royal Society of London, Robert Hooke declared, "The business and design of the Royal Society is to improve the knowledge of natural things, and all useful arts, manufactures, mechanic practices, engines and inventions, by experiment (not meddling with divinity, metaphysics, morals, politics, grammar, rhetoric or logic)."[1] Similarly, in his *History of the Royal Society* (1667), Thomas Sprat maintained that the investigation of nature "never separates us into mortal factions; . . . [it] gives us room to differ, without animosity; and permits us to raise contrary imaginations upon it, without any danger of a Civil War."[2]

Given the intense factionalism following the English Civil War, the Royal Society's decision not to meddle with politics was sensible. Those who practiced science could pursue politics separately if they wished. Isaac Newton, for instance, was simultaneously President of the Royal Society and Warden of the Mint. Once begun, the separation of science from politics became a tradition, as is shown by a later incident. In 1831 some of the new class of professional scientists, seeking political influence, created the British Association for the Advancement of Science to act as a pressure group for government aid and recognition. Almost at once scientists from the older universities moved in and depoliticized the association.

On the Continent, scientists expressed their political views more openly. During most of the eighteenth and nineteenth cen-

237

turies they tended to be more radical than their colleagues in other academic fields. I already have noted the anticlerical tendency of French scientists. The Revolution and the Napoleonic era encouraged them to hold similar radical views in politics. The elite of science students, the *polytechniciens*, repaid Napoleon's favor by fighting vigorously in defense of Paris in 1814 and by rallying to his side on his return from Elba.[3] With the restoration of the Bourbons, scientists as a whole came under suspicion, but within a few years they were politically active again. Since the Revolution French scientists have tended to be liberal and socialist in their views. They have cooperated readily with left-leaning governments, such as that of the Popular Front in the 1930s, but under other governments—at least since the Dreyfus affair, which polarized feeling—they have split into factions.[4]

In Germany the center of politics generally was further to the right. Nevertheless, during the nineteenth century the natural sciences tended to be more liberal than the humanities. In the rising science of physiology, especially, young men supported the liberal reform movement of the 1840s. Rudolf Virchow fought on the barricades in 1848, founded his own medicopolitical weekly, and became a member of the Democratic Congress of Berlin. Later in the century, however, scientists joined other academics in supporting the Wilhelminian regime, which financed science generously, recognized its cultural importance, and fostered its ties with industry. The antagonism between scientists in Berlin (who held the most prestigious jobs and tended to be more liberal in outlook) and scientists in the provinces reached a peak during the Weimar period, when scientists also were divided politically.[5]

In nineteenth-century Russia the natural sciences, chemistry especially, tended to be progressive and democratic, and science students were in the forefront of political unrest.[6] In the Soviet Union today, enormous pressure is put on social scientists to endorse the regime, but natural scientists are allowed more freedom. In fact, some of the leading dissenters have been natural scientists. Among others, however, biologist Zhores Medvedev and physicist Andrei Sakharov ("father" of the Russian H-bomb) were confined to mental hospitals for their opposition to the regime.[7]

238

How have governments dealt with science? Science first was officially recognized as a national asset with the founding of the British Royal Society (1662) and the French Académie des Sciences (1666). Both societies were encouraged by the state for the technological benefits they promised. Britain's Charles II, for instance, was pleased with the Royal Society's measurements of longitude at sea, but dismissed the weighing of air as a "childish diversion." Until World War I the British government preferred to let scientists organize themselves and find their own research funds. Thus in 1851 the Astronomer Royal declared: "In science, as well as in almost everything else, any national government inclines us to prefer voluntary associations of private persons to organizations of any kind dependent on the state."[8]

French science, on the other hand, was centralized and state-supported from the start for the purpose of enhancing the nation's culture and increasing its military and economic strength. The Académie des Sciences paid its members with government funds and commissioned a wide range of scientific and technological projects. (Contrast this munificence with the lot of the first British Astronomer Royal, John Flamsteed, who had to equip his observatory at his own expense, notwithstanding the national importance of his work.) The French Revolution almost completely destroyed the country's scientific institutions but then splendidly rebuilt them with the intention of creating a science that would actively contribute to the welfare of mankind everywhere. This step was decisive in shifting scientific leadership from England to France.

In the nineteenth century, German governments set up science departments in universities and encouraged them to compete with one another for faculty and students. The Germans funded science lavishly and actively backed its liaison with industry. Although an Englishman, William Perkin, created the first synthetic dye in 1856, it was the Germans who made a commercial success of it, because they had enough technologically minded industrialists willing to invest in a science-based industry and enough trained chemists to staff the factories. Thereafter the German chemical industry dominated the European market—so much so, in fact, that during World War I French uniforms were colored blue by a German-made dye! State support of education ensured a steady flow of scientific manpower to German industry. Between 1880 and 1900, for instance, the German government spent ten times more on education than did the British government.[9]

In the two great wars of the twentieth century, governments drastically increased their involvement in science. In World War I the belligerent powers set their chemists to work; in World War II they kept a wide range of scientists out of uniform to invent weapons and other products for the struggle. In World War I German scientists synthesized nitrates for fertilizers and explosives (thus beating the British navy's blockade on imports from Chile), and in World War II they invented rockets. In this war, too, British scientists invented radar, and scientists in the United States produced the atom bomb. World War II proved to the governments of industrialized nations that support for science could bring rapid, massive results.

In recent years governments have invested heavily in science, especially in areas of defense, space exploration, and public health. By orbiting Sputnik I in 1957, the Soviet government challenged American leadership in science. The United States responded by launching its own space program, appointing a special adviser to the President on science and technology, and creating various bodies to consult with the scientific community. All these events indicate that the more that governments have invested in science, the more they have influenced what scientific fields are explored, even if within those fields scientists are free to engage in their own research projects. Scientists in turn have sought to influence governments. So far, however, the division of science into separate disciplines, each with its own professional organization and research priorities, has prevented scientists from lobbying effectively as a united force.

As the United States government co-opted more scientists and disbursed more research money for weapons research, a number of scientists became concerned at the use of science for political policies they rejected. In December 1945 a group of scientists formed the Federation of Atomic Scientists (later called the Federation of American Scientists) to mobilize support for civilian control of the atomic energy program. This initiative was followed by others: protests against nuclear fallout, disarmament discussions at Pugwash conferences, the strike against the Vietnam war at the Massachusetts Institute of Technology, and vigorous lobbying against the antiballistic missile. Through such actions as these, American scientists demonstrated that scientific research now is bound so closely to the national welfare that it no longer should be pursued without regard for its potentially dangerous consequences. Scientists increasingly are acknowledging their obligation to society as well as to their profession. Various organizations have sprung up to criticize alleged misuses of science. These groups range from the staid Pugwash Association, run by older scientists using quiet diplomacy, to the radical Boston-based Science for the People.

Communist governments control science directly. In mainland China science has been encouraged and neglected in turns. It was encouraged in the mid 1950s, when China was emulating the Soviet Union, and again in the mid 1960s. It fell from favor during the Great Leap Forward of 1958 and then in the Cultural Revolution of the late '60s. The latter changed Chinese science in two main ways. Scientists were ordered to switch from basic to applied research. Physicists, for example, concentrated on such fields as lasers, plasmas, magnetic materials, and insulin (which the Chinese were first to synthesize). Universities and technical institutes were closed and scientific journals ceased publishing. At the same time education was reformed in order to avoid the "elitism" of Russian and Western scientists. Science had to be taught to the masses so as to create a new, scientific man. Early in the 1970s the universities and technical institutes reopened without exams and with the requirement that teachers and students "go down to the countryside" for long periods to work on farms or in factories, thus becoming "Red" or proletarian rather than "expert." Mere intellectuals were vilified as the "stinking ninth category" after eight other categories of "class enemy."

In 1977, under the Hua Kuo-Feng regime, science again was redirected. As part of a plan to modernize agriculture, industry, and defense by the year 2000, science and technology were given a new importance. The Chinese Academy of Sciences, previously denounced for elitism, was made a prime mover in this undertaking. The goal was to be "both Red and expert." Hundreds of technical journals resumed publication. Universities offered advanced courses,

241

insisting on exams. Lip service still was paid to the idea of "going down to the countryside," but few scientists have been seen doing it. The decision also was made to import foreign science selectively, while remaining basically self-reliant. Here Mao was quoted: "Our policy is to learn from the strong points of all nations."[10]

ECONOMICS

Big science is expensive. Only the superpowers can afford their own high-energy accelerators. Most Third World countries must put their limited funds into very few fields if they are to create satisfactory research traditions. But money alone is not enough. If science is not valued, nothing will be spent on it. The greatest wealth of Spain, shipped from the Americas, coincided with the nadir of its science and technology. In the United States in the nineteenth century, basic science was widely dismissed as a dilettante activity of no redeeming commercial value.

Before the nineteenth century most research was paid for privately. In Britain especially, scientists either were men of means or depended on private patrons. In the seventeenth century Robert Boyle, son of the Earl of Cork and a founder of the Royal Society, personally paid for the work of a number of other scientists. After losing his Cambridge professorship (for refusing to subscribe to a religious test act), the naturalist John Ray was supported by a wealthy friend. In the eighteenth century Antoine Lavoisier financed his chemical research partly out of inherited money and partly by working as a tax-collector. Early in the nineteenth century Pierre de Laplace and Claude Berthollet paid for much research done in their school at Arceuil. Darwin had a private income. The physiologist Claude Bernard first married a wealthy woman but then had to approach Napoleon III for money.

Later in the century, leading industrialists began to support science.[11] Private foundations were created, such as the Carnegie Institute, the Rockefeller Foundation, and the Institut International de Physique Solvay (established by the Belgian magnate Ernest Solvay). In the 1920s the Helmholtz-Gesellschaft (established by German in-

242

dustrialists to support physics) and a companion state foundation made an important innovation in research funding. Whereas previously foundations had decided what projects to support (American foundations continuing to do so until after World War II), the new foundations allowed panels of practicing scientists, chosen by members of a national scientific community, to allocate funds.[12]

As an example of the influence of private patronage on the course of research, take the financing of the infant science of molecular biology by the Rockefeller Foundation from 1932 to 1959.[13] Three factors in particular led the foundation to switch most of its support from physics to molecular biology: public rejection (resulting partly from the Depression) of the technology of the physical sciences; the appointment of Max Mason to the presidency of the foundation; and Mason's choice of Warren Weaver to direct the natural sciences program. Mason and Weaver, colleagues at Wisconsin, had discussed the possibility of applying recent insights and techniques in physics and chemistry to problems in physiology. Mason himself, watching his wife go incurably insane, wanted science to find a remedy for such illnesses, and he restrained Weaver from putting the Foundation's money into quantum mechanics. Weaver, ordinarily interested in statistics, then read up on genetics and biochemistry, and decided to support the research that was to lead 20 years later to the discovery of the structure of DNA. Without the backing of the Foundation, work on molecular biology hardly would have begun during the 1930s. The foundation financed the work of, among others, Herman Muller, Max Delbrück, Linus Pauling, and William Astbury.

The dependence of science on economics is equally evident in the Soviet Union. In 1917 the Bolsheviks seized control of a relatively undeveloped country. They realized that to build socialism, they had to transform a mainly peasant society into an industrial one. As Lenin put it, "Electrification precedes socialism." From the start the revolutionary leaders sought to harness science to military and economic needs. In 1929 Stalin made science serve Marxist theory. The truth of a scientific theory, he declared, lies in its economic consequences—science develops in response to economic needs. The Central Committee of the Party was made ultimately responsible for science, and under Stalin's leadership the Committee adopted Trofim Lysenko's genetics.

Lysenko promised that he would revolutionize Russian agriculture, which was lagging badly. Flatly rejecting both the neo-Darwinist view that acquired characteristics are never inherited and the Western practice of hybridization (cross-breeding of two existing strains), he maintained that environmental influences by themselves could alter the genes of crops to produce superior variations. Spring wheat, for instance, with high protein content and low resistance to cold, could be "transformed" by environmental conditioning into winter wheat, which has the opposite characteristics. Sprouts of spring wheat were exposed to low temperatures and then planted and harvested. Their offspring then were sown in the fall like winter wheat. Up to 99 percent died, but, according to Lysenko, those that survived differed physically and genetically from their predecessors. They had been changed into a winter wheat with a high protein content, capable, he said, of transmitting this content to its offspring.

Lysenko's thesis never was confirmed, but it sat well with the communist doctrine of human perfectibility in a socialist state, and so became the official line. Although Lysenkoism was a disaster to Soviet science and agriculture alike, its author was allowed to dominate genetics for more than two decades, not being dismissed until after the fall of Khruschev in 1965, when he was denounced as a fraud by a commission of top Soviet scientists. Eleven years later he died in relative obscurity. Yet his ideas still have life in them. At the Mironovka Research Institute of Selection and Wheat Breeding, they are quietly being put into practice, and may be due for a national revival (*Los Angeles Times*, December 5, 1977).

Another casualty of Stalinist science was Moscow's Medicogenetic Institute. In a study of the influences of heredity and environment, the institute had placed 1,000 sets of identical twins in selected environments. The findings were strongly for heredity. Infuriated by this affront to Party doctrine, Stalin abolished the institute and had the director shot.

Today in the Soviet Union the scientific community is recognized as the final judge of scientific truth, but Soviet scientists still are required to adhere to Marxian dialectical materialism.[14] The new

244

Soviet Constitution (1977) guarantees freedom of inquiry provided the results do not threaten the state.

SCIENCE'S INFLUENCE ON SOCIETY AND CULTURE

Science in turn acts on society and culture in three main ways: as a force of production (that is, as a body of knowledge furthering technological invention and the more efficient organization of work); as a method of inquiry and code of conduct applicable to other spheres of life; and as a source of ideas for other branches of knowledge.

A Force of Production. Scientific knowledge is used in various ways in the process of production. Through the application of particular laws, theories, and data, new technological devices and processes are invented. Scientific knowledge also is applied to production management, especially in automated factories.

In this century scientific knowledge has played an increasingly important role in promoting economic growth when compared with increases in the size of the labor force and the quantity of fixed capital. In the United States, during the two decades before 1929, increases in the labor supply and capital goods together accounted for about two-thirds of the increase in output, whereas in the decades 1929–1959 they were responsible for only 44 percent. Conversely, improved education and training, the source of only 13 percent of growth in the first period, doubled that figure in the second period. Finally, improved technology, resulting from the application of scientific knowledge, led to 12 percent and then to 20 percent of growth in the two periods.[15] Today the fastest growing industries—computers, electronics, optics, polymers—are based on extensive research in pure science. The computer and the laser, for instance, came out of the application of quantum theory to solid-state physics and optics.

A Method and an Ethos. Many scientists have proposed the scientific method and scientific conduct as ideals worthy of emula-

245

tion by society as a whole. The scientific method, they say, can be used more effectively than any other to regulate affairs in politics, commerce, and other spheres. The scientific community also is claimed to be a model of the good society, since the norms and values of science provide a moral discipline suitable for everyone. Let us look at some examples.

After the English Civil War in the mid seventeenth century, men of science saw themselves as particularly qualified to end political and sectarian strife. Claiming, as Robert Boyle said, to take "the whole of mankind for their care" and to put "narrow-mindedness out of countenance," [16] they proposed the method and morals of the new science as a model for a new social and political order. By imitating the cooperativeness of science, they declared, men would put an end to the factionalism that had caused the war; by subjecting their ideas to interpersonal criticism and verification, they would find policies on which all could agree.

Again, in mid-Victorian England a number of writers, such as Thomas Huxley, John Tyndall, Herbert Spencer, and Samuel Galton—some of whom were scientists, some not—sought to make science rather than religion the dominant force in the national culture. Huxley, the most effective publicist, claimed that all phenomena of life, mind, and society could be explained in terms of matter, motion, and evolution. Scientific knowledge and method, he said, could solve moral problems and even explain the meaning of birth and death. He called for a secular society in which scientists, not priests, would be the teachers. Since then, however, two world wars, using science for massive destruction, have cast doubt on Huxley's claim that science can provide an adequate moral foundation for human behavior. Ironically, it was Thomas Huxley's grandson, Aldous, who proposed the idea of a universal religion (based on features common to all religions) without a personal God.

In the United States, too, early in this century, many scientists and their supporters were convinced that only the method and moral code of science could restore rationality and probity to national life. In the words of a president of the National Academy of Sciences,[17]

It must be recognized more and more clearly that the scientific method is the one most likely to lead to results of permanent value.

246

. . . It is most desirable that our government should utilize to a greater and greater extent this method which is free from partisanship and has only truth to serve.

If the method and ideals of science were widely adopted, he continued, they would produce the moral and cultural regeneration necessary for lasting social reform. People would become like scientists: honest, forward-looking, and cooperative. Walter Lippmann agreed. Science, he declared, would provide not only "a binding passion, but a common discipline" to halt the disintegration of American democracy.[18] However, as we have seen, scientists at work are more partisan and self-interested than their apologists have allowed, and certainly no more moral than other professionals.

In many developing nations, Western science has been introduced simultaneously as a force of production, a moral example, and a method for the national regulation of affairs.[19] The hope has been to create through science a new culture and a new economy, enabling these nations to compete successfully in the modern world. However, instant modernization through science seems to have prospered only when subordinated to a broader program of social, cultural, and political reconstruction administered by a highly centralized authoritarian government. Otherwise, as in India, Western importations clash for the most part unavailingly with traditional attitudes and institutions.

Whatever their popular appeal, all such proposals—like the proposal of some Pugwash scientists to end Cold War politics with scientific discussion—overvalue the scientific method and ethos. The scientific method, as I have tried to show, is essentially a refinement of common sense, not a magical formula for solutions, and the scientific ethos is a set of norms which have less influence on the scientist than the cognitive assumptions of his research tradition. The scientific method is designed to make and to evaluate competing claims to knowledge, not to reach an accommodation between competing claims to power, which are the stuff of politics. The scientific ethos is too narrow to provide a morality for everyday life.

A Source of Ideas. Science also provides ideas for other fields of inquiry and especially for world views. We have seen how Darwinian evolution formed a biological foundation for the world view of late Victorian laissez-faire. We also have observed that Newtonian physics underlay deterministic materialism, the world view of most of the German middle class in the late nineteenth and early twentieth centuries. Supporters of the Enlightenment revered Newton for ordering the world. Wrote Alexander Pope as an epitaph on Newton:

> Nature and nature's laws lay hid in night.
> God said, let Newton be, and all was light.

The Newtonian order reinforced mid-Victorian optimism, although the idea of the "heat death" of the universe, predicted by the science of thermodynamics, contributed to the pessimism of some circles at the close of the century. (According to the second law of thermodynamics, the spontaneous dispersion of energy will cause the universe to cool irreversibly until eventually all heat is lost.)

In general the external impact of scientific ideas is greatest when the science is young and still in close touch with outside thinking. Contrast the theories of Copernicus and Einstein. Copernicus's heliocentric theory shook his contemporaries because it claimed to refute the Aristotelian theory that had been incorporated into the established religion. By subordinating the earth to the sun, it made man physically insignificant in a universe that was indifferent to him. Once accepted, Copernicus's theory changed the culture's world view. Man ceased to regard himself as a specially privileged being to whom the entire cosmos was sensitive, and saw himself instead as one who was able to comprehend a universe that dwarfed him physically.

The response to Copernicus can be traced in the poets.[20] At first men were disoriented by the loss of the old, man-centered universe. Shortly after Galileo had produced telescopic evidence of the "imperfection" of the heavens, the young John Donne—womanizer, intellectual, man about town—lamented:

> And new Philosophy cals all in doubt,
> The Element of fire is quite put out:
> The Sun is lost, and th'earth, and no man's wit
> Can well direct him, where to look for it.

248

As the new view grew familiar, however, dismay gave way to an intoxication with the idea of infinite space. The heavens now bore witness to a God of limitless power and creativity. Half a century after Donne, in his epic *Paradise Lost*, John Milton used the Ptolemaic scheme to show that for God the focus of the world was man. Yet this world has a Copernican vastness, which the fallen Satan is the first to experience. Accompanied by Sin, his paramour, and Death, his child, Satan looks out awestruck into chaos:

> Before their eyes in sudden view appear
> Th ιι ιι ι ι ι ιι ι ι ι ι ι ι ι ι, ι ι ιι.
> Illimitable Ocean without bound,
> Without dimension, where length, breadth,
>   and highth,
> And time and place are lost.

Einstein's special theory of relativity anticipated nuclear technology by proposing the equivalence of matter and energy. Nevertheless, neither the special nor the general theory had anything like the popular impact of Copernicus's theory. Great writers looked elsewhere for their ideas of order—James Joyce to the *Odyssey*, T. S. Eliot to the Church of England, Ezra Pound to Confucius. One of the few writers to use the scientific concept of relativity as a structuring device was the novelist Lawrence Durrell in the *Alexandria Quartet*. Both relativity and quantum mechanics (with its stress on the limits of observation) made some contribution to the relativism and skepticism common among liberal Western intellectuals since the end of World War I. However, the influence depended on muddled thinking, since neither theory says anything about human life or about thought in general. Heisenberg's uncertainty principle also has been cited as an argument for free will—again mistakenly, for the principle governs the motions of particles, not acts of choice.

Less developed than physics, biology has become more controversial. Man's physical insignificance in the cosmos has been accepted for some time, but his status among living things still is uncertain. Hence breakthroughs in biology are bound to stimulate vigorous, often indignant, responses from many quarters. Darwin's

249

theory of evolution raised a storm of criticism because it removed man from his primacy among living things and made him the chance result of a chance process of evolution. Darwin's theory not only challenged theology, it also revolutionized thinking in a range of intellectual fields; and it gave rise to a number of theories of social change collectively called Social Darwinism and epitomized in Herbert Spencer's theory of the upward march of history. However, the Social Darwinists oversimplified Darwin's ideas, ignored his denial that evolution represented progress, and mistakenly applied his biological theory directly to social affairs. They took for granted the stability of the economy within which the principle of natural selection, governing economic competition, operated (so they thought) to produce greater well-being. When the economy was seen to be distorted by monopolies, and governments began to regulate competition, Social Darwinism declined.

Today molecular biology is provoking new discussions of human nature. This science now has made it possible to alter human genes and so human beings. It also raises the crucial question whether, if all biology can be explained in physicochemical terms, there is any ultimate difference between human or any other life and inorganic matter. Theories influenced by the same world view tend to reinforce both one another and the world view itself. The physicalist research tradition of molecular biology is an extension of the mechanist world view still powerful in Western thinking. Mechanism maintains that human beings are determined either innately, as by their genes, or environmentally, as by natural selection or operant conditioning. Behaviorist theories, mechanistic in their assumptions, tend to take an optimistic view of man, since they take for granted that he can be changed—for the better. Nevertheless, according to behaviorism, a man does not change himself but is changed by external conditioning.[21] Hence molecular biology, Skinnerian behaviorism, Jensen's and Shockley's theories of the genetic determination of intelligence, and anthropological and ethological theories of man's innate aggressiveness (such as those of Robert Ardrey and Konrad Lorenz) all tend to support one another. They are opposed by schools of thought such as Freudian psychoanalysis, existentialism, and humanistic psychology, which affirm the individual's ultimate responsi-

250

bility for his behavior and, in differing ways, emphasize his capacity to change himself from within.

Specific findings in some of the life sciences, especially medicine, and their accompanying technologies directly affect people's health. Think of blood circulation, anesthesia, antibiotics, and the contraceptive pill. Hence they are quicker to influence people's attitudes and activities than corresponding achievements in the physical sciences. Studies of the brain, the heart, the reproductive organs, and the influence of genes on racial characteristics and intelligence, are immediately controversial because they concern subjects that are central to a person's self-image.

Consider, finally, how a conception of science as such forms part of the world view, not of a class or a country, but of an entire civilization. A major element in the world view of Western societies is the belief in progress through science, a belief first expressed by Francis Bacon. With his thesis that "knowledge is power"—roughly, that if you know nature and society, you can exploit the one and remake the other—Bacon linked progress in science with progress in general. Newtonian mechanics then showed what sort of knowledge could lead to what sort of power. Inanimate matter was reduced to systems of particles acting mechanically (that is, without purpose) under the influence of forces such as gravitation in accordance with Newton's laws of motion and the inverse square. During the eighteenth century, French scientists, notably Laplace, and their spokesmen, the Encyclopedists (Voltaire, Diderot, and others), argued that Newton's laws apply to all phenomena whatever, living and nonliving, and that the universe forms a single determined system of material parts interacting mechanically. But whereas Newton had believed that God still was active in His creation, later scientists maintained that after making the world, God had left it to run independently of Him. The world, they said, can be investigated as a working system and, when understood, it can be manipulated.

Scientific progress, then, becomes not only growth in reason but also growth in the power to invent new technologies and social techniques. Through better technology men improve their material condition, and through better techniques of social organization they improve the way they live and work together. Thus progress in

251

science leads to progress in technology, which, it is claimed, leads to material and moral progress.[22]

This notion of progress through science has guided and justified the advance of science since the second half of the eighteenth century. It has given rise to the conviction that science marches toward an ever more complete and accurate representation of what the universe is really like. It also is accompanied by an ideology. Whereas the scientific world view—more strictly the science-related part of our civilization's world view—states what is assumed to be the case and what is assumed to be possible, the ideology states what is assumed to be desirable and what it is assumed should be done. Where the world view may be expressed as a statement or set of statements to the effect that progress is possible through science and technology, the ideology may be expressed as a moral imperative, "Make progress, support science and technology!" This ideology has inspired the public awe of science (now tempered with an awareness of science's limitations), government patronage of science both direct and indirect (e.g., through education), the volume of scientific work, the size of the scientific profession, and the fusion of science and technology in industry.

IN RETROSPECT

How much influence should historians of science attribute to such external factors as I have mentioned here? According to one school of thought, the first task of scientific historiography is to disply the rationality of science, and therefore the historian should concentrate on factors internal to the research process, excluding the workings of the scientific community.[23] This school sees the history of science as basically a history of ideas developing from one to another with relatively little external influence. Where external influences exist, the history of science is that much less rational.

I believe this view to be mistaken in at least two respects. First, since the institutions of the scientific community are designed to promote research, their influence cannot be intrinsically irrational. Their procedures are rational in intent, if imperfect in practice. If a division is to be drawn between internal and external factors, it should be drawn between the scientific enterprise as a whole and the outside world rather than between the research process and the procedures of the community. Second, to treat only internal factors as intrinsically rational is to assume that all external factors have irrational effects without investigating whether they do in fact. Yet external factors often have stimulated scientific growth and therefore must be regarded as at least partly rational in their effects. Philosophy, religion, technology, social demand, patronage, and other external factors frequently have led scientists to ask questions and investigate phenomena that otherwise they might have ignored. Sometimes, to be sure, external factors have hindered science. It is the historian's duty to assess where they have been helpful and where not. This is not to say that internalist histories serve no useful purpose; the history of science may be written from many points of view, and it is as reasonable to focus on the ideas of science as on their social and economic context. Rather, it is to say that internalist histories are neither complete nor exclusively correct.

Other historians have focused on the sociocultural background of science. Sociologist Robert K. Merton has studied the influence of Puritanism on the physical sciences in seventeenth-century England and has analyzed present-day scientific institutions. Boris Hessen and John D. Bernal have investigated the part played in scientific development by social classes and economic forces. In my view, the chief flaw in such approaches is their tendency to underestimate the motive force of purely scientific ideas. Merton overplays the hold of Puritanism on English scientists because he pays too little attention to their intellectual concerns. Marxist historians such as Hessen and Bernal are too quick to read class interests into science. "Science," writes Hessen, "flourished step by step with the development and flourishing of the bourgeoisie. In order to develop its industry, the bourgeoisie needed science."[24] In fact, industry took little notice of science until the late nineteenth century.

What we need most are multicausal studies of the many different influences that at any one time bear on the mind of the scientist and the growth of science. The internal-external distinction is not entirely artificial, but it must be transcended if historians are to do justice to the variety of these influences. Recently some mainly internalist historians, such as Thomas S. Kuhn, have drawn attention to the effect of external factors on the choice of phenomena for investigation. But such studies do not go far enough. The objective of a truly multicausal history is to show how the various influences at work enter into the actual practice of research and into the substance of the scientist's conclusions. A start has been made in this direction by such works as Arnold Thackray's study of the influence of religion and other factors on Newtonian chemistry in eighteenth-century England, Robert Young's analysis of the relation of Darwinian evolution to the changing world view and ideology of nineteenth-century England, Paul Forman's investigations of the place of physics in Weimar Germany, and Gerald Holton's work on the interplay between the psychology of the scientist and the intellectual environment.[25]

It should be the aim of such studies, as it is an aim of this book, to show how the creative scientist acts, not as a technician isolated in a laboratory, but as a whole person facing the society and culture of his day, both choosing from them and formed by them, at once a child of his time and a contributor to it.

SUMMARY

Scientists have tended to avoid politics, although radical attitudes sometimes have appeared among Continental scientists and more recently among scientists the world over. In the past few years scientists have become more politically minded in response to growing government involvement in science, a process guided (say some) more by militarism than humanism.

Governments have supported science since the mid-seventeenth century, but they did not invest heavily in it until after World War II. Once dependent on private patronage, scientific research now is paid for mostly by governments, commercial interests, and foundations. Communist governments control science directly. In the U.S.S.R., Stalin put himself and the Central Committee of the Party in charge of science, and replaced Mendelian genetics with Lysenko's—a disastrous mistake. In China Mao shifted scientists from basic to applied research and sought to bring scientists and workers closer together.

Conversely, science influences the society and culture in which it is done. As a force of production, science is vital to the economies of the industrialized nations. The scientific method and ethos also have been proposed for adoption by the wider society. However, the scientific method is unsuited to politics, where values are in conflict rather than facts, although it occasionally may be applicable to the economics of production and distribution. The scientific ethos is too narrow a code for everyday life.

Science also is a source of ideas for other branches of inquiry and for world views. These ideas have their greatest impact when the science is young and most in touch with general movements of thought. Thus biology today is more controversial than physics, partly because it is a less developed science and partly because it bears more directly on human nature. The conception of a progressive science as itself a source of material and moral progress also is an important element in the world view of Western societies. We should bear in mind, however, that scientific findings in themselves do not morally obligate us to behave in certain ways rather than others.

Some historians of science concentrate on the internal movement of scientific ideas. Others focus on the sociocultural context of science. What we need are multicausal studies which trace the influence of a range of external factors on the actual conduct of research and on the conclusions reached by scientists. I cite a few such studies that have been published recently.

In my next chapter I will examine the interrelation between science and what is now the most important of the external factors with which it comes into play—technology.

NOTES

1. H. Lyons, *The Royal Society, 1660–1940: A History of Its Administration Under Its Charters* (London: Cambridge University Press, 1944), p. 41.

2. Thomas Sprat, *History of the Royal Society*, ed. with a critical apparatus by Jackson I. Cope and Harold Whitmore Jones (St. Louis, Mo.: Washington University Press, 1958), pp. 104–5.

3. Robert Fox, "Scientific Enterprise and the Patronage of Research in France 1800–70," *Minerva* 11 (1973): 466.

4. Joseph Ben-David, *The Scientist's Role in Society*, p. 106.

5. See Paul Forman, "The Financial Support and Political Alignment of Physicists in Weimar Germany," *Minerva* 12 (1974): 39–66.

6. Lewis S. Feuer, *The Conflict of Generations* (New York: Basic Books, 1969), p. 146.

7. Everett Carll Ladd, Jr., and Seymour Martin Lipset, "Politics of Academic Natural Scientists and Engineers," *Science* 176 (June 9, 1972): 1095.

8. George Biddell Airy, "Presidential Address," *Report of the British Association for the Advancement of Science* 20 (1851): li.

9. Hilary Rose and Steven Rose, *Science and Society* (Baltimore: Penguin, 1970), p. 29.

10. See Deborah Shapley, "China after Mao: Science Seeks To Be Both Red and Expert," *Science* 197 (August 19, 1977): 739–41; *Science News* 112 (August 13, 1977); and "C.P.C. Central Committee Circular on Holding National Science Conference (September 18, 1977)," *Peking Review* 40 (September 30, 1977): 6–11.

11. For an inventory of academic physics around 1900, see Paul Forman, John L. Heilbron, and Spencer Weart, *Physics circa 1900: Personnel, Funding, and Productivity of the Academic Establishments*, Historical Studies in the Physical Sciences, ed. Russell McCormmach, vol. 5 (Princeton: Princeton University Press, 1975).

12. Paul Forman, "Financial Support and Political Alignment of Physicists in Weimar Germany," 50–51.

13. Robert Olby, *The Path to the Double Helix*, foreword by Francis Crick (Seattle: University of Washington Press, 1974), pp. 440–43. See also Robert E. Kohler, "The Management of Science: The Experience of Warren Weaver and the Rockefeller Foundation Programme in Molecular Biology," *Minerva* 14 (Autumn 1976): 279–306.

14. See Loren R. Graham, *Science and Philosophy in the Soviet Union* (New York: Knopf, 1972). Relations between the Party and the Soviet Academy of Sciences had deteriorated so far that the celebrations marking the Academy's 250th anniversary

(1974) had to be postponed. The following year the main speaker on the Academy's election day was Mikhail Suslov, chief Party ideologist, who admonished the scientists to devote their efforts to "the cause of the Party and the people." *Science News* 108 (December 6, 1975). Most recently (July, 1977) Benjamin Levich, a specialist in physicochemical hydrodynamics, was denied permission to attend an international conference held in his honor at Oxford University on the grounds that he possessed state secrets and had "besmirched his country to the detriment of his scientific work." On the other hand, physicist Benjamin Fain was granted an exit visa (after three years) to assume a post at Tel Aviv University.

15. See Edward Denison, *Sources of Economic Growth in the United States* (New York: Committee for Economic Development, 1962).

16. Thomas Birch, ed., *The Works of the Honourable Robert Boyle* (London, 1774), p. 20.

17. Ira Remson, "Opening Address," *Science* 37 (1913): 722.

18. Walter Lippmann, *Drift and Mastery* (New York: Kennerley, 1914), p. 154.

19. In 1964 Harold Wilson, then in opposition, forecast that a new Britain would be "forged in the white heat of this [scientific] revolution" to be created by his new plan for science.

20. See Margaret M. Byard, "Poetic Responses to the Copernican Revolution," *Scientific American* 236 (June 1977): 121–29.

21. True, Skinner holds that a person can condition himself, e.g., by rote cultivation of habits. Nevertheless, such a person does not change from within but imposes changes on himself, the inner self remaining unaltered. A person changes consciously from within by performing for their own sake acts that he believes to be morally desirable, or personally liberating, or both. Unlike habits mechanically cultivated, these acts draw on the instinctual energies at the core of the psyche and so permit that core to be changed.

22. On the rise of this world view, see Henryk Skolimowski, "The Scientific World View and the Illusions of Progress," *Social Research* 41 (Spring 1974): 52–82; and Hans Jonas, "The Scientific and Technological Revolution: Their History and Meaning," *Philosophy Today* 15 (Summer 1971): 76–101.

23. E.g., Alexandre Koyré, *Metaphysics and Measurement: Essays in Scientific Revolution* (Cambridge: Harvard University Press, 1968); Charles C. Gillispie, *The Edge of Objectivity* (Princeton, N.J.: Princeton University Press, 1960); A. Rupert Hall, *From Galileo to Newton, 1630–1720* (London: Collins, 1963); Imre Lakatos, "History of Science and Its Rational Reconstructions," in *PSA 1970: In Memory of Rudolf Carnap*, ed. Roger C. Buck and Robert S. Cohen, Boston Studies in the Philosophy of Science, vol. 8 (Dordrecht, Holland: Reidel, 1971), pp. 91–136.

I distinguish between the research process and the procedures of the scientific community. Similarly Stephen Toulmin distinguishes between the discipline and the profession of science. See his *Human Understanding*, vol. 1 (Princeton: Princeton University Press, 1972).

24. In N. I. Bukharin et al., *Science at the Cross Roads: Papers Presented to the International Congress of the History of Science and Technology, Held in London, June 29 to July 3, 1931, by the Delegates of the U.S.S.R.*, 2d ed., with a new foreword by Joseph Needham and a new introduction by P. G. Werskey (London: Cass, 1971), p. 170.

25. Arnold Thackray, *Atoms and Powers;* Robert Young, "The Historiographic and Ideological Contexts of the Nineteenth Century Debate on Man's Place in Nature," in *Changing Perspectives in the History of Science*, ed. Mikuláš Teich and Robert Young, pp. 340–438; Paul Forman, "Weimar Culture, Causality, and Quantum Theory, 1918–1927: Adaptation by German Physicists and Mathematicians to a Hostile Intellectual Environment," in *Historical Studies in the Physical Sciences*, ed. Russell McCormmach, vol. 3, pp. 1–115; Gerald Holton, *Thematic Origins of Scientific Thought.*

Chapter 11

Science and Technology

WHAT IS TECHNOLOGY?

The word "technology" is derived from the Greek noun, *techne*, meaning art or skill. This derivation tells us that technology essentially is a practical endeavor, one of altering the world rather than understanding it.[1] Where science pursues truth, technology prizes efficiency.[2] Whereas science seeks to formulate the laws obeyed by nature, technology uses these formulations to create implements and apparatus that will make nature obey man. Like science, however, technology is an enormously complex entity consisting of phenomena of many kinds—agents, institutions, products, knowledge, techniques, and so forth. Here I consider it as a historically developing enterprise for constructing machines and other artifacts, devising techniques and processes, transforming and creating materials, and organizing work, so as to satisfy human wants.

The chief aim of technology is to increase the efficiency of human activity in all spheres, including that of production. Technology produces more varied objects to meet a wider range of wants, and it improves particular types of objects to meet particular wants more fully. Technology improves objects by making them, for example, longer-lasting, or more reliable, or more sensitive, or faster in performance, or a combination of these, depending on the object's function. Thus, reinforced concrete lasts longer than brick, the Mount Palomar telescope is far more sensitive than Galileo's, and today's computer calculates infinitely more rapidly than the abacus. Technology also improves production by reducing the time or cost of making a certain object, as by manufacturing a particular material, telescope, or computer more quickly or less expensively than another of the same kind.

259

Technological work is purposive and rational. The technologist visualizes an end, draws up a plan to attain it, and employs tools, materials, and techniques to execute the plan. His immediate goal is to make something or bring something to pass; his long-term goal is to realize some further state of affairs through what he has made or brought to pass. When an engineer builds a bridge, he forms an idea of the kind of bridge that will suit a particular site. He then turns this idea into a detailed plan or design, and finally supervises the construction of the bridge according to the plan. His immediate goal is to construct the bridge. His long-term goal is to make possible a certain flow of traffic across the bridge.

Aside from the routine exercise of skills, one can achieve technological results only by skillful planning and careful use of tools. In all complex cases, technological action involves theoretical as well as practical reasoning, systematic knowledge as well as expertise. This was recognized in classical antiquity. Aristotle defined techne as "identical with a state of capacity to make, involving a true course of reasoning."[3] In the Middle Ages, Hugh of Saint Victor called technology "a form of knowledge which must embrace the methods of production of all things."[4] At that time such technologies as alchemy and architecture rested on their own theoretically ordered knowledge. Since the nineteenth century, however, many technologies have based much of their knowledge on the theories and findings of science.

How do technologists think? Thomas Edison called invention 1 percent genius and 99 percent sweat. Another inventor, Elmer Ambrose Sperry, said that invention is 110 percent sweat. What goes on during the sweat? In some ways technological thought resembles the concrete, visual thinking of the arts. This resemblance was pointed out by Aristotle. Art and technology, he said, create things which did not exist before, whereas science and philosophy deal with things which exist naturally or by necessity.[5] Technological thought also resembles the practical reasoning of common sense, as was pointed out by the historian of science, Alexandre Koyré. "The technical thought of common sense," he wrote, "does not depend on scientific thought, of which it nevertheless can absorb the elements, incorporating them into common sense."[6]

Art and practical reasoning are manifest both in the design of a product and in the finished work. A design not only is a representation of the structure of something to be created, it also is a plan of action for creating it. The final product has a practical function, but it also is potentially a work of art, since it is man made, having a certain form and sometimes an aesthetic finish. Cars and computers, telephones and typewriters are made not only to relieve man of physical and mental labor but also to please the eye.[7]

SCIENCE AND TECHNOLOGY IN HISTORY: THREE VIEWS

How have science and technology been related in the course of history? According to one view, the most important innovations in Western technology, especially since the seventeenth century, have rested on laws, theories, or data established by pure science. This view was especially popular in the nineteenth century. In 1832 Joseph Henry, lecturing at the Albany Institute, New York, declared that "every mechanic art is based upon some principle or general laws of nature and . . . the more intimately acquainted we are with these laws, the more capable we must be to advance and improve the useful arts." Thus, said Henry, James Watt invented his steam engine by using Joseph Black's theory of latent heat; shipbuilders employed Euler's mathematical studies of the curvature of hulls, and Humphry Davy invented the mine safety lamp after he had studied fire-damp scientifically. Likewise Robert Fulton's achievements in steam navigation and Eli Whitney's invention of the cotton gin "depended on their extensive scientific knowledge."[8] Similar opinions were held in England. In 1835 Charles Babbage wrote: "It is impossible not to perceive that the arts and manufactures of the country are intimately connected with the progress of the severer sciences; and that, as we advance in the career of improvement, every step requires, for its success, that this connexion should be rendered more intimate."[9]

According to another view, the decisive partner has been technology. Granted that today important advances in technology depend

261

on fundamental scientific research, nevertheless this research is carried out in the first instance to provide the knowledge needed for those advances. It is technological needs that give vigor and direction to fundamental scientific research.

This view has been advocated by Marxist historians. For example, in 1931 the Soviet historian Boris Hessen proposed that Newton's *Principia* "in the full sense of the word is a survey and systematic resolution of all the main group of physical problems" confronting transportation, communication, industry, and warfare at the time. According to Hessen, the rising middle class needed science to develop the forces of production. For this purpose four main classes of problems, all in mechanics, needed to be solved: simple machines, sloping surfaces, and statistics; ballistics; hydrostatics, hydrodynamics, and atmospheric pressure; celestial mechanics and tidal motion. The first group involved mining and construction work; the second, artillery; the third, mining, smelting, canal-building, and shipbuilding; the fourth, navigation. Newton, writes Hessen, "in the center of the physical and technical problems and interests of his time," constructed a theory which unified celestial and terrestrial mechanics and at the same time helped solve these problems. The first book of the *Principia* expounded the general principles used in doing so. The second book, dealing with the motions of terrestrial bodies, solved problems in the first three classes. The third book, treating celestial and tidal motions, solved the last group of problems.[10]

Today the most common view is that science and technology developed for the most part independently of each other until about 100 years ago. According to the historian of science, A. Rupert Hall, "virtually all the techniques of civilization up to a couple of hundred years ago were the work of men as uneducated as they were anonymous." He and Marie Boas Hall declare that "the beginnings of modern technology in the so-called Industrial Revolution of the 18th and early 19th century owed virtually nothing to science, and everything to the fruition of the tradition of craft invention." The inventions of the Industrial Revolution were "the results of empirical experiments, products of craft skill and large quantities of hard labor."[11] It was not until the latter half of the nineteenth century that the first industrial laboratories were set up and businessmen began to employ scientists to invent and improve the relevant technology. Before this, most technological advances were the work of inventors and craftsmen using practical know-how and little or no theoretical science. For example, the supreme engineering achievement of the Renaissance, the printing press, was a product of many craftsmanly skills, such as paper-making, ink-making, metallurgy, block printing, printing with metal type, and the production of the screw press itself.

On this view, the Industrial Revolution was set off by social and economic factors rather than by science. The steam engine, which used the natural forces of heat and steam to power machines for manufacturing and transporting goods, was invented by skillful entrepreneurs who employed scientific methods only occasionally and had little scientific knowledge. Basic science was used only once—by James Watt when he invented the separate condenser in 1764. Watt had been an assistant to the physicist Joseph Black at Glasgow University. After studying Black's data on latent heat, he realized how important it was to avoid the enormous loss of heat in heating and cooling the cylinder of the existing steam engine. Watt's successors in the pioneer steam industry were mainly practical mechanics and engineers with no training in physics or mathematics. Nevertheless, the steam engine showed what technology might accomplish if it applied theoretical scientific knowledge rather than the data and working principles of craftsmen and engineers.

According to the Halls and others, technology first made significant use of science in the late nineteenth century, when the chemical industry drew on scientific findings, at first to alter natural substances, as in the dye, fertilizer, and pharmaceutical industries, and eventually, to synthesize new substances altogether by reorganizing the molecules of existing ones. In the laboratories of the chemical industry the technological potential of the experimental method was realized for the first time, and nature was manipulated not for intellectual but for practical purposes. The trigger event occurred in 1856. William Henry Perkin, an 18-year-old assistant at the Royal College of Chemistry in London, tried to synthesize the drug quinine and got an unexpected result. He found that the mauve dye produced by his first attempt at synthesis would neither wash from cloth nor fade after prolonged exposure to light. He immediately realized that the demand for the dye would be enormous if only it could be produced more cheaply than natural dyes. Nevertheless, it was German firms, employing professional chemists, who exploited the discovery most successfully.

In the 1880s electrical power technology emerged with the construction of the first public power stations. This work drew on Faraday's law of induction (that the motion of a magnet near a wire "induces" an electric current), proposed in 1831. At the end of the century Maxwell's electromagnetic theory led to the invention of the radio. Unlike the chemical industry, which emerged from prescientific crafts, the electrical industry provided power for the propulsion of machines by means of an invisible force which science itself had enabled human beings to control. Before the rise of industrial chemistry, man had worked and refined natural substances, as in pottery and metallurgy, but science alone enabled him to manipulate electricity. World War II produced the technology of electronics. Unlike previous technologies, which had satisfied mainly biological needs (as for food, shelter, clothing, and locomotion), electronics met a need created by civilization itself, the need to process and transmit information.

Nevertheless, it is unrealistic to suppose that science and technology remained entirely separate so long. Their interactions were less frequent, certainly, but also more fluid and complex, and hence are less easily detected.[12] Before the late nineteenth century, in my view, science and technology were only relatively independent of each other, and since then the alliance between them has been less complete than the orthodox view suggests. Indeed, the alliance was not fully formed until after World War II. The early automobile and aircraft industries, for instance, owed little to science.

Science often has been stimulated by technological problems and inventions. In antiquity the lever and the balance preceded the laws by which they were explained. In the Renaissance the Venice dockyard aroused Galileo's interest in mechanics. William Gilbert, the first European to study electricity and magnetism scientifically, learned much from seamen about the magnetic compass. Evangelista Torricelli's refutation of the thesis that nature abhors a vacuum arose from attempts to explain the limited capacity of pumps used in draining mines. In the seventeenth century the emerging science of chemistry took much of its information and apparatus from alchemical practice. In the eighteenth century, problems in mining stimulated the birth of geology. Lavoisier's study of niter and oxygen was assisted by his work for the government's gunpowder monopoly. In the nineteenth century Carnot and Clapeyron used the work of steam engine designers and Darwin that of stock breeders. James Prescott Joule, codiscoverer of the first law of thermodynamics, turned to the study of heat while attempting to design an electromagnetic engine.[13] In the United States the science of ecology began in an attempt to understand the spread of thistles and other immigrant weeds in midwestern grasslands.[14]

Science, in turn, has been applied to technological invention and improvement in more subtle and varied ways than is yet realized. More science than we think was used in the early Industrial Revolution. In France, for example, the chemist Claude Berthollet, a director of the national dye industry, discovered that the gas chlorine, isolated earlier by the Swede Carl Scheele, could readily bleach cotton fabrics. Nicolas Leblanc, a physician, found that soda could be made from salt and sulfuric acid. Both products then were manufactured. Indeed, throughout the eighteenth century the French government enthusiastically promoted the application of science to technological undertakings.[15] In England the potter Josiah Wedgwood, guided by scientific knowledge, conducted experiments for many years to produce an extraordinary number of new glazes. Before the Industrial Revolution, science contributed to mining (through geology), medicine (through biology), and agriculture.[16] It should be pointed out, too, that science supplied technology not only with specific findings but also with the scientific method of investigation, certain laboratory techniques, and a belief in the usefulness of research.[17]

SCIENCE AND TECHNOLOGY TODAY

Science currently is regarded as the partner of technology and, in this respect, as a utilitarian as well as a contemplative enterprise. Material progress is made by continually constructing new wealth- and efficiency-producing devices that are constructed and operate according to scientific laws and theories. Progress in technology is held both to justify progress in science, which makes technological progress possible, and to provide visible evidence of it. Thus the continuing intellectual progress of mankind includes, and the continuing material progress of mankind requires, the progress alike of science and technology.

Modern science-based technology is the use of pure and applied science to build artifacts, construct techniques, and organize human activities. Electronics technology, for example, is the study and use of knowledge about electron absorption and emission—knowledge largely produced by the applied science of electronics—to construct devices that will transmit images (television), record and reproduce sound (tape recorders and high fidelity systems), and store and handle information (computers). The products of technology, which include artifacts such as space satellites and cameras, techniques such as systems analysis, and forms of organization such as the assembly line and the bomber fleet, embody many findings of both pure and applied science.

A distinction also should be drawn between research and development. Research is an activity which produces knowledge. It therefore includes pure and applied science together with the research done by technologists. Development is the use of knowledge to design new products, create prototypes, and tool up for production. It normally excludes production itself, routine product testing, and quality control.

Modern technology not only produces machines and physical tools but also organizes and systematizes activities. Physical ("hard") technology draws on the natural sciences, nonphysical ("soft") technology on the behavioral ones. Thus an organization or management system which incorporates principles of scientific management is as much a technological construct as is a moon rocket. Automotive technology, for example, includes not only the machinery used on the assembly line but also the organization of the factory and the industry. Techniques which embody behavioral science findings include public opinion polls, market surveys, educational tests, programs for computer-assisted instruction, and urban-planning procedures. Indeed, in some areas more work goes into nonphysical technology than into its counterpart. For example, far more time and money now are spent devising instructional programs for education than constructing the hardware to convey them. Here, however, I am concerned with physical ("hard") technology.

An extraordinary number of modern technological products and processes rely on knowledge and instruments created by basic research. Let me mention some technological windfalls from physics alone. Modern media of communication operate through the transmission of electromagnetic waves, whose existence was predicted by Maxwell and confirmed experimentally by Hertz. Induction coils in automobiles observe Faraday's law of induction. Nuclear power was made possible by the discovery and splitting of the atomic nucleus by Rutherford, Fermi, the Curies, and others. Basic circuits used in computers were discovered by nuclear physicists in the 1930s. Transistors were invented by physicists doing quantum mechanical research into solids. The laser, the vacuum tube, the Josephson junction, and the superconducting magnet all were invented in the course of basic research.

(The *laser* provides a light beam intense enough to vaporize the hardest materials. It has been used to drill holes in diamonds, weld the retina of an eye to its supports, and perform microsurgery on parts of single cells. The basic research for the laser and the *maser*—a similar device operating in the microwave or radio part of the electromagnetic spectrum—was done by Charles H. Townes and his colleagues at Columbia University in the 1950s. The *vacuum tube*, a basic electronics component now partly replaced by the transistor, is an evacuated glass or metal envelope in which electricity is conducted through a gas or vacuum. The *Josephson junction* is a thin insulator separating two superconducting materials, i.e., metals (such as lead or tin) that lose all electrical resistance at temperatures near absolute zero. It can be used to measure very weak magnetic fields. The Josephson effect, the flow of an electric current across such an insulator, was predicted in 1962 by the British physicist Brian Josephson. The *superconducting magnet* is a powerful electromagnet made from a superconducting material.)

267

None of these technological applications were foreseen by the scientists themselves. Indeed, it is not always possible to anticipate the eventual technological applications and social consequences of a basic research project at the time it is begun. When Maxwell proposed his theory of the electromagnetic field, neither he nor his contemporaries had radio or television in mind. When Einstein advanced the theory of special relativity, neither he nor anyone else thought of a military or commercial use for atomic fission. As late as the 1930s Rutherford dismissed the possibility as "moonshine." Like some other scientists, he thought that splitting the atom would require as much energy as it would yield. Not so many years ago a British Astronomer Royal derided the notion of space travel as "bilge."[18]

Basic and applied science tend to flow into each other, just as applied science merges with technology. In physics most applied research now is being done in the study of condensed matter and especially in metal, high-polymer, thermal, and solid-state physics. Much of atomic, molecular, and electron physics, and also of plasma physics and the physics of fluids, consists of applied work. Acoustics and optics have been regarded as applied branches for some years, because their fundamental theories have long been established and are used to investigate particular phenomena of potential practical significance.

Sometimes fundamental discoveries opening up new fields in basic science are made in the course of applied research. Both cosmic background microwave radiation and radiation from space in the radio spectrum were discovered by scientists at Bell Telephone Laboratories who were trying to solve practical communications problems caused by interference in radio and microwave communication. For this practical purpose it was sufficient to measure the frequency, intensity, and directionality of the noise and correlate it with environmental factors. But the scientists were talented enough to see the theoretical significance of their discoveries, which then was explored by basic research.

The interaction between science and technology can be illustrated from the history of nuclear physics, applied nuclear physics, and nuclear engineeering.[19] Until the atom was split at the end of the 1930s, the interest in nuclear physics was mainly intellectual. Physicists sought to observe regularities in nuclear structure and reactions and to discover the ultimate nature of nuclear forces. But as soon as fission was discovered and the notion of the neutron chain reaction was formulated, much existing basic knowledge at once became applicable. Many experiments and theories that had been developed sufficiently for intellectual purposes now needed to be made more precise and comprehensive to meet practical needs. Thus applied research was in fact done in nuclear physics, although at the time it was not classified as such. Then the growing use of radioisotopes and the creation of nuclear power reaction had become obligational for nuclear physicists to develop them technologically. New problem areas opened up in neutron physics, nuclear chemistry, and radiation biology, and in the effects of pile radiations on the materials used in the construction of reactors. More reliable techniques had to be devised for counting particles. The properties of radionuclides had to be studied and measured more carefully. In its early stages little of this work was important enough intellectually to the specialty of nuclear physics to justify the money and talent invested in it (though later it stimulated such fundamental achievements as the theory of nuclear shell structure). Instead the work was justified by its practical value. One cannot, then, fix an exact point at which nuclear physics turned into applied nuclear physics, or just when applied nuclear science became the technology of nuclear engineering.

Basic research often is done with a twofold motive. A problem is chosen for investigation because the solution will add to theoretical understanding in the specialty and also because it may prove technologically applicable. In applied research a problem is selected not only because a solution is needed for some practical purpose but also because basic science has advanced to the point where the problem is ripe for solution. Thus intellectual opportunity and technological need coincide in the same problem. It is precisely this choice of problems with a joint intellectual and practical significance that distinguishes applied science from technology. The technologist in a hurry need only solve the problem in the particular case. For him knowledge is the means to a technological solution. The applied scientist, on the other hand, seeks a full understanding of all aspects of a problem, and hence a solution that applies to a range of cases. The applied scientist may or may not take longer to find a solution, but if he does, it will have many more technological applications and will save time and energy in the long run. Finally, we should bear in mind that a fair amount of work in applied science and technology is done independently of work in basic science. Applied science also may investigate problems posed by technology, and technology may draw on practical experience and inventiveness.

Finally, there is a return flow of information, problems, and instruments from technology to basic science. The expansion of nuclear weapons and computers stimulated the further growth of nuclear physics, which had made them possible in the first place. The development of the transistor contributed to the growth of solid-state physics; work on the gas laser, to atomic spectroscopy; and the development of the superconducting magnet, to low-temperature physics. Modern physics owes to technology such instruments as the Geiger counter, amplifying systems, vacuum equipment, the electron microscope, the wind tunnel, and the particle accelerator. There have been similar feedbacks to chemistry and biology.

IS TECHNOLOGY AUTONOMOUS?

I turn now from technology, the partner of science, to technology as a social and cultural force.

It is sometimes said, especially by certain economists, that technology is neutral and value free, being merely a response to economic demand. A demand arises for a product to fill a need, and the technologist designs the product for the sake of the money that he or someone else will make from it.

This picture is grossly oversimplified. Today, in industrialized nations, demands arise to fill wants as much as needs, and wants are stimulated by a host of factors, not the least of which are advertising campaigns designed to implant them. Technologists in turn are moved not only by the profit motive but also by humanitarian, intellectual, aesthetic, and purely personal motives, such as a delight in invention itself. To make something new, especially if this is difficult or dangerous, is an irresistible challenge, and an ingenious solution to a technical problem gives aesthetic satisfaction. Of his response to a clever recording instrument, Frank Tatnall (inventor of the strain gauge) wrote, "I get a secret feeling of elation whenever I see an automatic time axis provided without any mechanism to work it, nothing added to size, weight, or cost."[20]

Far from being the neutral result of market forces, technology has a character of its own. In differing degrees, and always in conjunction with other forces, it can constrain the technologist and affect the values and lives of those who use it. Since it satisfies a want, an effective technological innovation usually expresses some previously held value of a culture, yet through its concrete form it can develop or distort this value. Think of the clock, the steam engine, the assembly line, and the computer. In the working of these inventions men have seen an image of their lives, and the inventions themselves have provided the means to make this image a reality.

When clocks were installed in the marketplaces of Western Europe, the clocks, not the authorities, led people to think of time as an ever-circling wheel carrying them forward minute by minute toward eternity. People began to regulate their lives mechanically, not because they were told to, but because the means for doing so stared them in the face.[21]

The steam locomotive, with its continuous conflict between untamed energy and purposeful discipline, seemed to represent the necessary tension of human life—the tension of Faustian man or the nineteenth century entrepreneur, to whom life was a never-ending struggle. This spectacle of energy under restraint may well have contributed to the creation and popularity of the Marxist notion of dialectical tension and the Freudian conception of instinctual repression. It entered everyday speech in the form of such colloquialisms as "working under pressure," "blowing up," and "letting off steam."[22] As a means of rapid transportation, the locomotive also helped people lead the high-pressure lives they thought appropriate.

Today computers encourage people to mechanize their notions of mental activity. Because the computer can perform functions analogous to those of the mind, there is a tendency to think of the mind *as* a computer. The mind is "programmed," memory is "information

271

retrieval," problem-solutions are "read-outs." When the mind is so computerized, any free, spontaneous action comes to be regarded as an abandonment to instinct rather than as an integrated behavior of the whole person in which reason and passion coincide.

Technology not only affects the consumer, it also makes demands on those who invent and exploit it, for it has an internal dynamic. When a new technique has been invented, its users tend to seek further applications and to expand the organization they have created. Thus, once the SST (supersonic transport) program got under way, it was hard to stop. Again, when a businessman has invested heavily in production machinery, he normally is committed to making further investments, not only to sell the product but to cope with subsequent contingencies (such as the need to redesign the product in response to competition or faulty performance). Further, once research has started, it sometimes leads to inventions of no particular interest to the company. Technology is not a neutral agency that can be started and stopped at will; it has an inner drive of its own.

Technology both opens doors and closes them. On the one hand, it enables people to do things not otherwise possible. On the other, it pushes people to act for technical rather than humane reasons. When people enter through one door, opened by a technical innovation, they may find that another door closes. The wisdom of a particular choice may not become clear until many years later when, for example, the range of options has been narrowed by the consequences of the original decision. Over half a century ago, public authorities in this country decided to subsidize highway construction to aid automobile transportation without at the same time improving other modes of transportation. The automobile increased individual mobility and hence personal freedom. But, in addition to taking its own death toll, it was a prime cause of suburban sprawl, inner city decay, and air pollution.

Large technological undertakings also produce their own bureaucracies. Such undertakings tend to create technical elites, centralize control, and make decisions that reflect self-interest; they rarely limit or regulate themselves.[23] Because economic growth or military strength depends on their work, technical elites also can muster wide support for their projects. And because a scientific and technological outlook predominates in the culture of industrial societies, public authorities who finance technology tend to make their decisions in the spirit of that outlook, supported by the advice of those very elites.[24]

Although most technological undertakings acquire a momentum of their own, in the sense that they have technological or social consequences not immediately foreseen, technology itself is not really autonomous. There is no underlying technological force bent on its own course, carrying empires and societies with it. Technology is in the hands of its creators and operators, not the reverse. Technological innovations are the work of people who, as a rule can persuade society to adopt their product only by showing that it promotes some value the society prizes.[25] Once an innovation is accepted, it generally becomes more widely diffused, and its social consequences often stimulate further technological innovation. During the first half of this century the airplane was of serious interest mainly to one group in society, the military. It was only when the Atlantic could be crossed nonstop by high-speed jets, and when the price and accident rate of air travel were sharply cut, that people were persuaded to fly in large numbers.

Other considerations tell against technological determinism. For one thing, most important technological advances—and not just those in weapons and space exploration—now result from government decisions to finance certain projects. For another, a given innovation may be differently used and have different effects in other lands and cultures. In the United States the radio led to a greater interest in politics and made politicians more sensitive to public opinion; in some other countries it was used to increase totalitarian control and weaken public opinion. Or contrast the attitudes to technology taken in classical China and Western Europe. By 1500 A.D. the Chinese had mastered bronze and iron casting, invented gunpowder, and sailed their junks beyond the China seas. Yet none of these achievements in China led to a mercantile capitalist society or to a longing for world domination. Instead of exploiting the power their technology gave them, the Chinese used it for limited ends.

273

ATTITUDES TO TECHNOLOGY

What do people think of technology? We can distinguish three main views.[26] The first, and by far the oldest, regards technology as an unmitigated evil. In many legends of the Golden Age, whether set in the past, as by Genesis, or in the future, as by utopias, men and women go naked, eat raw fruit and vegetables, and live the most primitive lives. In these legends the two fundamental technological innovations—the use of fire to cook and of animal skins for clothing—are said to result from man's degeneration. They are punishments, not achievements. According to this view, the more man advances technologically, the more he corrupts himself.

Opposed to this is the view, so well expressed by Francis Bacon, that only through technology can man regain the happiness and sovereignty over nature that he had before the Fall. According to Bacon, nature exists for man's sake. "The whole world," he wrote, "works together in the service of man; and there is nothing from which he does not derive use and fruit . . . insomuch that all things seem to be going about man's business and not their own." To make nature serve his purposes, man must know her laws, and he will discover them not in the wisdom of the ancients ("fruitful of controversies but barren of works") but through empirical science. Bacon called for a gigantic, state-aided research program to uncover the secrets of nature and create a vast new technology "for the relief of man's estate." In his utopia, *The New Atlantis* (1627), Bacon described a society governed by an order of philosophers, the Merchants of Light, who were "the noblest foundation that ever was upon the earth." "The end of our foundation," declared their spokesman, "is the knowledge of causes and the secret motions of things; and the enlarging of the bounds of human empire, to the effecting of all things possible." Bacon's dream of continual progress achieved through science and technology harmonized with the traditional Christian belief that God had created nature for man's use.[27]

Around 1830 the Baconian ideal was updated and propagated enthusiastically by the disciples of Saint-Simon. Mankind, they said, was about to turn from warfare to industry and from competition to cooperation. "The exploitation of man by man has come to its end.

274

. . . The exploitation of the globe, of external nature, becomes henceforth the sole end of man's physical activity."[28] The London Exposition of 1851, with its masses of machinery housed in the dazzling new Crystal Palace, seemed a dramatic realization of the Baconian dream. To many, industry and laissez-faire economics promised a swift fulfillment of the Utilitarian ideal of "the greatest happiness of the greatest number." In his novel *Looking Backward* the American utopian Edward Bellamy foresaw in the year 2000 a planned society based on a beneficent technology. Throughout the first half of the twentieth century the Baconian belief was held more fervently than ever.

But the pessimistic attitude never died. Rousseau called for a return to small, technologically simple communities in which man would find moral purity. Romantic poets like William Blake denounced the "dark satanic mills" planted in "England's green and pleasant land." In her tale of Frankenstein, Mary Shelley expressed the fear that technology might destroy its creators. Ralph Waldo Emerson warned that "things are in the saddle and ride mankind." At the Paris Exhibition of 1900 Henry Adams recoiled before the power of the dynamo, "a symbol of infinity," and was appalled by the "destructiveness" of the automobile. In the early twentieth century, however, such pronouncements generally were regarded as atavistic. Most people believed that it was man's duty to transform nature for his own good, and that technology was the means of doing so.

In the 1960s the pessimistic attitude made a comeback. Aldous Huxley's *Brave New World* (1932) portrayed a society in which man had been debilitated by a technology giving him every bodily comfort. Rachel Carson's *Silent Spring* (1962) revealed the destruction of plant and animal life by the unrestricted use of pesticides. Jacques Ellul's *La Technique* (1954) argued that man had absorbed technology so deeply that he had become spiritually and physically dominated by it. Today many people are horrified by a technology which produces nuclear weapons that threaten the survival of the race, drugs that mutilate human beings, electronic devices that invade freedom and privacy, and automobiles that ruin the quality of life. Science fiction, a bellwether for the popular mood, constantly creates dystopias in which men are controlled more effectively by electronic devices than by brute force and deception. Technological growth has become suspect not only to writers and laymen but also to many scientists and technologists.

Some of these pessimistic views refute themselves, however. They overvalue nature and undervalue man. Barry Commoner's "third law of ecology," for instance, states that "nature knows best." But surely he must know that nature possesses neither knowledge nor mentality. Natural processes take their own course indifferent to the human condition. Over millennia a complex balance of process has established itself. Whenever people introduce a new machine, or drug, or variety of plant, they tend to modify that balance. Inevitably, then, they risk worsening their state. But risk is inherent in action, for one can never be sure that what he does will not have at least some harmful consequences. The sound course is not to rely on some presumed wisdom of the cosmos but to increase our knowledge so that we may better anticipate the consequences of our actions.

It also has been proposed that if we are to preserve the biosphere and the life it supports (including our own), we must return to the primitive view that nature is sacred. This is the view of nostalgic people who regard nature as a wise mother grieving for man, her erring child. It would be nearer the truth to see man himself as having the biosphere in his care. The biosphere is fragile, and we can damage it irretrievably. Forewarned, we are forearmed. We do not and can not dominate nature. The most we can expect is to achieve some measure of control by studying nature and cooperating with it.

Between the optimists and the pessimists are the moderates. According to them, both extremes seize on one potential of technology and amplify it to the exclusion of the other. What is needed is a cooler, less extravagant conception of technology, with its vast range of consequences, its great powers for good and ill, and its reflection of the complexity of human nature. In fact, the consequences of technological innovation almost always are more complicated than the innovators expected.

The moderates are right in saying that we cannot reap the benefits of technology without incurring some risks. If we upset the balance of nature, it cannot always be in our favor. Instead of abandoning technology to eliminate the risks, let us keep it and reduce the risks. It often may be difficult to reach agreement on what risks are acceptable, chiefly because variables are hard to quantify and conflicting values are involved. A proposed new airport, for instance, means noise and nuisance to some people, faster travel and more business to others. Even so, the attempt must be made. Reaching agreement means finding a middle way. It is unreasonable, for example, to require the Mississippi River to be so free from pollution that we can drink from it. We should be satisfied if fish can live in it. It also is unreasonable to say with Paul Ehrlich that "people are pollution." Instead of zero population growth or unrestrained population growth, let us have a population growing within limits. And let us tackle the problems to which such growth gives rise.

The conflict between the three views is exemplified in the current debate on economic growth. Optimists maintain that society should give priority to economic growth, even at some cost to social welfare.[29] Pessimists hold that if growth ends, social problems will disappear. Indeed they say, unless sweeping moral and political reforms are enacted, growth will stop anyway within the next hundred years, for nonrenewable resources will be depleted, food supplies will run out, and pollution will cover the globe.[30] But they underestimate the potential of technology, the regulatory power of governments, and people's capacity to learn from experience and change their behavior when faced with danger.

The moderates rightly point out that the crucial issue is not whether to grow or not grow but how to redirect economic output so that it meets the needs of mankind more effectively. Thus Ronald G. Ridker proposes that instead of attacking economic or population growth as such we should attack the particular problems they create.[31] He maintains for instance, that, even with low growth, pollution will increase substantially by the year 2000 unless strong countermeasures are taken. However, if the standards that have been recommended by the Environmental Protection Agency for 1976 are enforced by 2000, pollution will be cut well below present levels even with a high growth rate, and at a cost of less than 2 percent of the Gross National Product. In short, it makes better sense to stop pollution than to stop growth. Should we restrict growth and technological development because we are unsure of their effects? No, says Ridker, because then we shall never discover those beneficial effects through which we may improve the lives of future generations as well as avoid disaster.

TACKLING THE CONSEQUENCES

What are some of the effects of technology? It is strange how many well-intended innovations have had uneven consequences. On the one hand, medicine has doubled the average life span, wiped out many diseases, abolished many forms of pain, and provided the means of effective birth control. On the other, in controlling deaths it has created a population explosion, which continues because the means of birth control are largely ignored. In agriculture the "green revolution" makes it possible to abolish starvation, yet food is not distributed evenly, and many people are starving to death. In transportation, construction, and power supply, inventions have improved the quality of life, yet power production and the internal combustion engine pollute both air and water. Nuclear power makes a nuclear holocaust possible, but it can provide energy when fossil fuels run out. The computer threatens to turn into "big brother," yet it helps us regulate complex social processes. Molecular biology holds out the hope of abolishing disease and creating a permanent food supply, yet it could some day be used to breed an elite of evil geniuses and a mob of brutish followers.

How are we to cope with these effects? First we must recognize that some of them are more social and political than technological. In such cases we may know what the technological effects are and what means can be used to avoid them. What we do not know is how to persuade people to adopt these means. We know, for example, how to end traffic congestion, urban decay, and most forms of pollution, but not how to convince people to drive less, rehabilitate neighborhoods, and use trash bins. All that scientists and technologists can do is to proclaim as loudly as they can what the consequences of certain kinds of behavior are likely to be.

But technology also has many harmful effects whose causes

278

and remedies are insufficiently known. These include the effect of industrialization on health, the production of carbon dioxide, and long-range influences on climate and atmosphere. Technological growth has interfered with an enormously complicated natural environment, a vast web of physical, chemical, and biological causes and effects, many of them little understood. Basic science has the formidable task of observing and explaining the interrelations among phenomena in this environment, of systematically tracing their causes and effects so as to produce knowledge undistorted by serving particular interests—knowledge that can be used to devise technical solutions for the problems that greed, ignorance, and technology itself have created.

We also must take action to anticipate, control, or forestall the consequences of technology before they occur and, if necessary, to mitigate them once they have occurred. A most important form of action is technology assessment, or the evaluation of the effects, both beneficial and harmful, of technological innovation. Technological assessment asks such questions as these: How will the proposed product or process affect human health, safety, and the quality of life? What effects may it have on the environment? Will it lead to more employment or less? How will it affect population movement? What balance can be struck between different, competing considerations?

Technology assessment often leads to bitter disputes among scientists, and between scientists and policy makers.[32] Think of the controversies over nuclear fallout from weapons tests, the use of herbicides and defoliants in the Vietnam War, the effects of DDT and other pesticides, the antiballistic missile, the supersonic transport, and the dangers of nuclear power plants. If wise policy decisions are to be made on issues arousing strong feelings and involving powerful interests, there must be fair hearings and due process, and the public must have access to all relevant information.

Nevertheless, we should be aware of the human frailties involved when expert advice is given on public issues. In principle it is possible to distinguish between technical questions and questions of policy. The former are those to which a determinate answer can be given, if not now, then at some future time. Such questions are answered scientifically. The latter are questions about who should pay for and who should benefit from particular actions. These questions involve value judgments and are answered by policy choices reached through the political process. While technical questions are answered best by experts, all persons have an equal right to be heard on questions of policy. An expert photochemist is the best judge of how many cases of skin cancer would be caused by a fleet of SSTs, but on the question whether the benefits of using the SST are worth the costs of skin cancer his opinion should carry no more weight than anyone else's.

In practice, however, the distinction between technical and policy questions often is ignored. The experts who advised the congressional committee on the SST did so with the intention of supporting or opposing the program, and they sought to influence policy. The economists from the Federal Aviation Administration who told the committee that the program provided 150,000 jobs did not also state that a similar number of jobs would be created by an equivalent federal expenditure on other projects. Scientists who said that a 1 percent reduction in stratospheric ozone would produce 10,000 cases of skin cancer per year did not also mention that, for the individual, the increase in ultraviolet exposure was equal to spending no more than a few minutes in the sun every now and then.

Again, the experts often fail to give definite answers or else give answers that are definite but mistaken. As Niels Bohr remarked, "It is difficult to predict, especially the future." In the 10 years between the announcement by President Kennedy of the SST program and the final congresssional hearings, scientists and technologists were unable to say definitely either that the SST would pay its way or that it would adversely affect the world's climate. Again, expert advice often is so technical that those who must make the policy decisions do not understand it, while policy makers who already have committed themselves seldom are swayed by expert advice. In the SST debate, politicians were moved more by public than by expert opinion. Moreover, the best qualified experts often are not asked for their advice, either because little effort is made to find them, or because their opinions are known but not wanted, or because recourse is had to a select coterie. Finally, expert advice may be obtained to support prechosen policies, either by the experts themselves, or by government agencies, or by politicians. This may be done by calling on experts who will support your case, by invoking more experts than your opponents, by giving advice in colorful and dramatic terms, by emphasizing extreme cases, and by other devious means.

In my view, most flaws in the advisory process can in principle be corrected. Further study of this process can produce insights that will lead experts to give better advice and politicians to make better use of it. The social problems created by technology can only be solved politically. We cannot rely on the individual conscience. We all know that cars pollute, yet which of us will give up his car for that reason? Each knows how little his self-denial would contribute to the end of pollution.

How, then, and by whom is technology to be controlled? The best way would be through the transformation of social institutions so that responsibility and authority are spread widely through the citizenry.[33] But how in practical terms is this reform to be accomplished? Again, thinkers and politicians have barely begun to reflect on the need for a policy to direct technology in its entirety, a policy that would provide solutions to the problems we have mentioned, of nuclear technology, conservation, pollution, population growth, urban blight, species preservation, and genetic engineering. Each of these problems, difficult enough in itself, becomes more difficult internationally. A single nation cannot hope to keep its air and water clean, for they may be polluted elsewhere. One country cannot control the world's birth rate. International problems require international solutions. Yet few nations will be any happier to limit their populations than they are to restrict their armaments.

To find optimum solutions we need, on the one hand, a theory of science and technology as social forces and institutions and, on the other, a theory of the good society which science and technology could help us achieve. As a prerequisite, however, we need more public debate. Unless there is open, informed discussion of different technological alternatives, a satisfactory choice cannot be made between them. So far such alternatives have not found vigorous political advocates.[34]

Informed debate depends on a knowledge of the facts. Scien-

281

tists and technologists have a special responsibility to contribute to public enlightenment. They should communicate their findings on the actual and potential consequences of technology to the widest audience possible and in terms readily understood by the ordinary citizen. Important findings should be analyzed critically and controversially through all available media. If technology is to be used creatively to benefit mankind as a whole, we shall need an enlightened public ready to evaluate it fairly—something we do not have at present.

SUMMARY

Technology is the historically developing enterprise of constructing artifacts and organizing work to meet human wants. It has affinities with both art and practical action. It aims to increase the efficiency of human action in all spheres. In pursuit of this aim, technology draws on both practical and theoretical knowledge, and calculates the most effective means to given ends. During the last century it was widely believed that technological innovation depended on scientific knowledge. In our own century some historians have declared the opposite, that science grows in response to technological needs. Most historians, however, maintain that science and technology remained separate until German chemical firms began to employ professional chemists about a hundred years ago. Nevertheless, this view is mistaken. Some sciences, such as chemistry and geology, actually grew out of crafts, while certain products, such as soda and glazes, had been improved scientifically before the Industrial Revolution.

Today science and technology are interdependent. From science, technology derives basic knowledge, instruments, and techniques. From technology, science receives instruments and problems for solution. Science and technology interact in the realm of applied science, which is the investigation of problems whose solutions are expected to be technologically applicable.

Technology both expresses and develops existing cultural values. It even patterns the lives and values of its users, as in the case of the clock, the steam engine, the assembly line, and the computer. Through its internal dynamic, technology makes demands on its developers. For the organization of its larger projects it creates bureaucracies. Technology enables people to do things they could not have done otherwise, although particular technological choices inevitably foreclose others. (A society which turns from public transportation to the private automobile will find it hard to turn back.) Nevertheless, technology as such is not autonomous; it is created by human beings and is subordinate in the main to cultural values and government decisions.

Throughout history, technology has been feared as an arrogant departure from the state of nature, a fear that has been revived by the relentless expansion of technology since World War II. Francis Bacon, however, argued that through science and technology man could regain the sovereignty over nature that he possessed in Eden, and Western society has come increasingly to believe in the possibility of unlimited material and moral progress with the aid of science and technology. Nevertheless, neither optimism nor pessimism is fully justified. The wiser course is to aim for limited progress and keep its inevitable costs to a minimum.

Some technological innovation is essential and desirable. It has been necessary for the modernization of all societies, and it will enable our own to survive and improve. The development of new technologies must be encouraged, and the training of imaginative technologists fostered.

At the same time the many consequences of new technologies need to be critically assessed before they are introduced, and they must be monitored continuously thereafter. To this end, steps must be taken to correct weaknesses in the science advisory process, such as the tendency of experts to conflate advice and advocacy. Many technically brilliant schemes will have to be rejected because they seem likely in the long run to do more harm than good. Some will have to be canceled at the research stage before pilot projects have been given a run, lest strong pressures build to carry them to completion.

Technology will not be brought under social control until political parties propose specific policies for doing so. To hasten this process, there should be the widest possible debate and diffusion of information.

Every human faculty can be misused. Men can use their intelligence to enslave one another, their imagination to deceive, their eloquence to betray. But if they did not use these faculties, where would they be? By using his powers, man continually extends the range of his achievements. Technology increases his ability to do so.

Technology can create or destroy, make man more human or less. But civilizations, like individuals, must risk if they are to grow. If we exercise prudence to minimize the damage of technology and courage to maximize its benefits, surely the risk is worth taking.

NOTES

1. For the general reader there is some useful information on technology since the Industrial Revolution in Charles Susskind's *Understanding Technology* (Baltimore: Johns Hopkins University Press, 1973). A readable history of technology is D. S. L. Cardwell's *Turning Points in Western Technology: A Study of Technology, Science, and History* (New York: Neale Watson Academic Publications, Science History Publications, 1972). On the philosophic understanding of technology see Friedrich Rapp, ed., *Contributions to a Philosophy of Technology* (Dordrecht, Holland: Reidel, 1974); and *Technology and Culture* 7 (Summer 1966), "Toward a Philosophy of Technology." A good introduction to the moral problems of technology is Hans Jonas's *Philosophical Essays: From Ancient Creed to Technological Man* (Englewood Cliffs, N.J.: Prentice-Hall, 1974), Part I: "Science, Technology, and Ethics."

2. See Henryk Skolimowski, "The Structure of Thinking in Technology," in *Contributions to a Philosophy of Technology*, ed. Rapp, pp. 77–78: "Science aims at enlarging our knowledge through devising better and better theories; technology aims at creating new artifacts through devising means of increasing effectiveness."

3. Richard McKeon, ed., *Introduction to Aristotle* (New York: Modern Library 1947), p. 427.

4. Quoted by Friedrich Klemm, *A History of Western Technology*, trans. Dorothea W. Singer (Cambridge: MIT Press, 1964), p. 72.

5. McKeon, *Introduction to Aristotle*, pp. 427–28.

6. "Du monde de l'à peu près à l'univers de la précision," *Critique* 4 (1948): 809. Quoted by Edwin T. Layton, Jr., "Technology as Knowledge," *Technology and Culture* 15 (1974): 36.

7. See Cyril Stanley Smith, "Art, Technology, and Science: Notes on their Historical Interaction," *Technology and Culture* 11 (1974): 36.

8. Quoted by Arthur P. Molella and Nathan Reingold, "Theorists and Ingenious Mechanics. Joseph Henry Defines Science," *Science Studies* 3 (1973): 331–32.

9. Quoted by Peter Mathias, "Who Unbound Prometheus? Science and Technical Change, 1600–1800," in *Science and Society 1600–1800*, ed. Peter Mathias (Cambridge: Cambridge University Press, 1972), p. 60.

10. Boris Hessen, "The Social and Economic Roots of Newton's 'Principia,' " in *Science at the Cross Roads: Papers Presented to the International Congress of the History of Science and Technology, Held in London, June 29 to July 3, 1931, by the Delegates of the U.S.S.R.*, 2d ed., with a new foreword by Joseph Needham and a new introduction by P. G. Werskey (London: Cass, 1971), pp. 155–76.

11. See A. Rupert Hall and Marie Boas Hall, *A Brief History of Science* (New York: New American Library, 1964), p. 219. The Halls admit, however, that in "the chemical industries from the last years of the 18th century . . . scientific knowledge was conspicuous" (*ibid.*). Eric Ashby is even more skeptical of the scientific contribution to technology until late in the nineteenth century. In his view, "the Industrial Revolution was accomplished by hard heads and clever fingers . . . by men with no systematic education in science or technology. . . . There was practically no exchange of ideas between the scientists and the designers of the industrial processes." *Technology and Academics: An Essay on Universities and the Scientific Revolution* (London: Macmillan, 1958), pp. 50–51.

12. For this view, see A. E. Musson and Eric Robinson, *Science and Technology in the Industrial Revolution* (Manchester: University of Manchester Press, 1969); Peter Mathias, "Who Unbound Prometheus?" pp. 54–80; and some of the essays in Mikuláš Teich and Robert Young, eds., *Changing Perspectives in the History of Science*, especially Neil McKendrick, "The Role of Science in the Industrial Revolution: A Study of Josiah Wedgwood as a Scientist and Industrial Chemist," pp. 274–319.

13. Donald Cardwell, "Science and Technology: The Work of James Prescott Joule," *Technology and Culture* 14 (1976): 674–87.

14. Ronald Tobey, "Theoretical Science and Technology in American Ecology," *Technology and Culture* 14 (1976): 718–28.

15. See Roger Hahn, *The Anatomy of a Scientific Institution: The Paris Academy of Sciences, 1666–1803* (Berkeley and Los Angeles: University of California Press, 1971).

16. Mathias, however, contests this. "Who Unbound Prometheus?" p. 71.

17. See McKendrick, "Role of Science in the Industrial Revolution," p. 313, and Mathias, "Who Unbound Prometheus?" p. 79.

18. Bernard Dixon, *What Is Science For?* (New York: Harper & Row, 1973), p. 210.

19. Physics Survey Committee, National Research Council, *Physics in Perspective*, vol. 1 (Washington, D.C.: National Academy of Sciences, 1972), pp. 497–98.

20. Frank Tatnall, *Tatnall on Testing* (Metals Park, Ohio: American Society for Metals, 1966), p. 49. Quoted by Eugene S. Ferguson, "Toward a Discipline of the History of Technology," *Technology and Culture* 15 (1974): 23–6.

21. Lewis Mumford, *Technics and Civilization* (New York: Harcourt, Brace, 1934), pp. 12–18.

22. D. O. Edge, "Technological Metaphor," in *Meaning and Control: Essays in Social Aspects of Science and Technology*, ed. D. O. Edge and J. N. Wolfe (London: Tavistock Publications, 1973), p. 36. On the cultural impact of the railroad, see Leo Marx, *The Machine in the Garden: Technology and the Pastoral Ideal in America* (New York: Oxford University Press, 1964); Herbert L. Sussman, *Victorians and the Machine* (Cambridge: Harvard University Press, 1968); and T. R. West, *Flesh of Steel* (Nashville: Vanderbilt University Press, 1967).

23. John McDermott, "Technology: The Opiate of the Intellectuals," *New York Review of Books*, July 31, 1969, 25–35. Whether this analysis applies to big technological ventures before the Industrial Revolution remains to be shown.

24. Tom Bottomore, "Machines Without a Cause," *New York Review of Books*, November 4, 1971, 18.

25. John G. Burke, "Comment: The Complex Nature of Explanations in the Historiography of Technology," *Technology and Culture* 11 (1970): 23–24.

26. For a similar triad, see Emmanuel G. Mesthene, *Technological Change: Its Impact on Man and Society* (New York: New American Library, Mentor, 1970), pp. 15–20.

27. On the development of the Baconian ideal, and on some twentieth-century criticisms of it, see William Leiss, *The Domination of Nature* (New York: Braziller, 1972). On the influence of Judeo-Christianity on Western man's attitude to nature, see Lynn White, Jr., *Medieval Technology and Social Change* (Oxford: Clarendon, 1962), and "The Historical Roots of Our Ecological Crisis," *Science* 155 (March 10, 1967): 1205 ff., reprinted in the same author's *Machina ex Deo: Essays in the Dynamism of Western Culture* (Cambridge: MIT Press, 1968).

28. *Doctrine Saint-Simonienne: Exposition* (Paris: Librairie Nouvelle, 1854), p. 436. Quoted in Leiss, *Domination of Nature*, p. 82.

29. E.g., Wilfred Beckerman, *Two Cheers for the Affluent Society: A Spirited Defense of Economic Growth* (New York: St. Martin's, 1975).

30. Mihajlo Mesarovic and Eduard Pestel, *Mankind at the Turning Point: The Second Report to the Club of Rome* (New York: Dutton/Reader's Digest, 1975).

31. See *Scientific American* 230, 2 (February 1974): 42.

32. Some recent studies are those of Ian D. Clark, "Expert Advice in the Controversy About Supersonic Transport in the United States," *Minerva* 12 (1974): 416–32; Allan Mazur, "Disputes Between Experts," *Minerva* 11 (1973): 243–62; Dorothy Nelkin, "The Political Impact of Technical Expertise," *Social Studies of Science* 5 (1975): 35–54.

33. See Leiss, *Domination of Nature*, p. 197.

34. In 1964, declares Nigel Calder, "the science programs of the Republican Party of the USA and the Communist Party of the USSR were virtually identical. . . . If politicians had different views of how the nation and world should be, they should surely differ in the uses they wished to make of science." *Technopolis: Social Control and the Uses of Science* (New York: Simon and Schuster, 1971), pp. 279–80.

Chapter 12

The Scientist's Responsibility

So far I have considered the interactions of science and the wider world of man in a largely factual way. I have sought to describe what these interactions are rather than to judge how beneficial they are. I now want to examine two issues raised by this interplay: (1) Is science morally neutral? and (2) Should science be planned? The first issue arises from the influence of the outer world on science. If class interest, say, or financial patronage affects the course of research, can science seek truth for its own sake? The second issue is created by the impact of science on the world. If scientists produce knowledge that may be either beneficial or harmful, should they alone decide what knowledge to seek? Both issues concern the scientist's relation to the truths he claims to seek.[1]

SCIENTIFIC TRUTH AND THE NORMS OF SCIENCE

The scientist, we have seen, seeks the truth about the order of nature. In doing so, he accepts a stringent discipline. He follows certain standards of conduct or "norms" which in effect state what he should do to attain the truth. As the means to an end valued for its own sake, these norms are moral values; observing them constitutes scientific integrity. For example, one scientist may compete with another to make a discovery, but he must not falsify his results. A scientist may work on his own, but he knows that he is part of a much broader community dedicated to discovering the truth. The value of every conjecture, every test, every observation report he makes rests on his commitment to this end.

Granted, a few scientists have betrayed it. In 1974 William A. Summerlin of the Sloan-Kettering Cancer Society in New York City

288

admitted that he had painted the feet of mice to make it appear that they had successfully received skin grafts from nonidentical twins. A genuine grafting would have broken the immunological barrier that causes transplants to be rejected. Again, a young researcher, Robert Gullis, wrote to the British journal *Nature* confessing that data he had published describing the effects of certain opiates on nerve-like cells were completely fraudulent, "mere figments of my imagination." "I was so convinced of my ideas," he stated, "that I simply put them down on paper." One of his co-authors, his senior at the Max Planck Institute in West Germany, was staggered. Gullis should have known, he said, that "lies in science have short legs." They cannot outrun the truth.

Or take the case of the midwife toad. In the early 1920s Paul Kammerer, a pillar of Viennese biology, believed fervently in Larmarck's theory of evolution. The theory states, against Darwin, that characteristics acquired by an organism in its lifetime may be inherited by its descendants. Kammerer proposed to prove the theory as follows. Most toads mate in water, the male gripping the female's slippery body with dark "nuptial pads" on his feet. If, therefore, Kammerer forced midwife toads (which normally mate on land) to copulate in water for several generations, he would expect the males to develop the pads too. In fact, he announced that they had done so. But either he or an assistant had simulated the pads with india ink. Unaware of the fraud, Soviet scientists, likewise Lamarckians (see chapter 10), offered Kammerer a post at Moscow University. Kammerer accepted, but the fraud was discovered, and Kammerer killed himself.[2]

Nevertheless, such flagrant violations are rare. The rigorous crosschecking of the scientist's results by his peers effectively deters them. More to the point is the inevitable fact that not all scientists live up to their ideals all the time. Again, the exception proves the rule. Between 1913 and 1917, in a celebrated series of experiments the American physicist Robert Millikan measured the charge of the electron. In a paper published in 1910, describing some prior experiments, Millikan wrote: "I have discarded one uncertain and unduplicated observation apparently upon a singly charged drop which gave a value of the charge of the drop some 30 percent lower than the final value of *e*." About Millikan's "great scientific honesty" Paul Dirac had this to say:

> Many experimentalists when they get a result which is against what they are trying to establish and against the whole mass of their other observations would say simply "My apparatus was out of order on that day. I don't know just what was wrong, but I cannot duplicate the result, so it is not of any interest to think of it further and it is not worth mentioning it when I come to publish my work." Well, Millikan was not like that. He was scrupulously honest and had to refer to experiments which were discordant with what he was trying to prove.[3]

Ironically there is no agreement on the norms of science. Many lists and definitions have been proposed, but none has won general acceptance. Jacob Bronowski, for instance, argued that scientists "*ought* to act in such a way that what *is* true can be verified to be so." If the aim of science is "to explore the truth," then scientists must be "independent" as individuals and "tolerant" as a group. From these values there follow a range of others: "freedom of thought and speech, justice, honor, human dignity, and self-respect."[4] Yet justice, honor, dignity, and self-respect are associated not with science alone but with any morally worthwhile human activity, while the values of individualism and tolerance, which Bronowski considers primary, actually presuppose those he considers secondary. So let us consider sociologist Robert K. Merton's list, which has been the most widely discussed.

The Mertonian Thesis. According to Merton, the scientist's quest for truth implies four norms: universalism, communality, disinterestedness, and skepticism.[5] To these norms Merton and his followers later added those of originality and individualism.[6]

By *universalism* Merton means that any claim to knowledge should be evaluated regardless of the status of the scientist making it and regardless of external considerations such as any economic, political, or even humanitarian consequences likely to follow from it. *Communality* means that any piece of scientific knowledge belongs to the scientific community rather than to the scientist who produced it; *disinterestedness*, that the scientist should not do research solely to enhance his professional reputation, social prestige, or income; *skepticism*, that he should take no claim to knowledge on trust; *originality*, that he should think originally; and *individualism*, that he should be free to choose his research problems and techniques, and evaluate his findings, without interference from anyone in authority.

According to the Mertonians, those norms are ideals in the minds of scientists. Scientists use them to guide their own behavior and assess that of colleagues. The norms are related to the goals respectively of science (the growth of knowledge) and the scientist (professional recognition) through the reward mechanism. Thus, the scientist is expected to produce original knowledge, for which recognition is given, yet he also must make his contributions common property. Since he normally is denied all rights to his property except that of being recognized as its originator, he values recognition the more highly.

Although professional recognition is a powerful incentive, the norm of disinterestedness dissuades the scientist from admitting this openly. Informal pressures in fact require him to regard the welfare of science as more important than personal gain. They require him rigorously to examine all claims to knowledge, both his own and those of other scientists. Moreover, he must be willing to accept responsibility for any inadequacies found in his published papers. Observing all these requirements, says Merton, maximizes scientific productivity and minimizes the probability of undetected error.

Critique of Merton. Merton's view is open to several objections. Scientists simply do not observe his norms. Contrary to the norm of disinterestedness, scientists are highly competitive. And contrary to the norm of originality—contrary, indeed, to the whole set of norms whose raison d'être is to stimulate scientific growth—most major theories initially met with indifference or contempt. Copernicus's heliocentric theory was not generally accepted until a century after his death (though it was used at once to make predictions). William Gilbert's studies of magnetism and electricity were dismissed by Francis Bacon as "so many fables." Thomas Young's wave theory of light was scornfully rejected by the Newtonian establishment. Probably no work was more vilified at first than Darwin's *The Origin of Species.* An anonymous critic in the *Edinburgh Quarterly Review* debunked it as a "rotten fabric of guess and speculation . . . dishonorable to natural science." Darwin's former geology professor wrote to him from Cambridge, "I laughed . . . till my sides were almost sore . . . utterly false and mischievous . . . deep in the mire of folly." Pasteur's theory of fermentation was spurned by chemists, and his germ theory was denounced by physicians. Quantum theory was ignored for over a decade and would have been ignored longer had not Poincaré recognized its merit. And so it goes. This sad chronicle led Max Planck to declare, not altogether fairly, that "a new scientific truth does not triumph by convincing its opponents and making them see the light but rather because its opponents eventually die and a new generation grows up that is familiar with it."[7]

Moreover, it often is the most original scientist who breaks these norms. Some of the most creative scientists are notoriously sensitive about their reputation and skeptical of most claims to knowledge but their own. It may be replied that exceptions can be made for geniuses. But Kuhnian normal scientists also are exceptions since they are neither original, nor skeptical, nor individualistic. Nor do applied scientists observe the norms. In government and industry, scientists are neither disinterested nor individualistic but work for pay on problems chosen by their employers. Yet again, the scientist usually is committed more to his research tradition than to a set of norms. Research traditions are characterized by their cognitive assumptions rather than by norms of conduct. The values they emphasize are intellectual, and they vary from one tradition to another.[8]

Merton, it is true, admits that in practice the norms are violated. Perhaps, then, they are ideals honored more in the breach than in the observance. Yet it is doubtful whether there is now, or ever was, normative consensus among scientists. One observer has written, "If there ever were a reasonably firm consensus with respect to scientific values, it was not maintained long after 1920. It may never have been more than a myth."[9] Some scientists, for example, defend secrecy as a means of preventing other scientists from stealing their results. Others defend a degree of dogmatism, arguing that if we took seriously every challenge to accepted knowledge, we would give the claims of cranks and charlatans precedence over worthwhile research.

One might argue, however, that even if the Mertonian norms are neither practiced nor professed, they still are the ideal norms for science. The fact that they are ignored does not make them any less justified. But how ideal are they? The norm of universalism requires the scientist to observe only scientific considerations in conducting and judging research. Yet scientists acknowledge a higher moral claim whenever they refuse to perform certain kinds of experiments such as those on live human beings. Again, recognizing that social influence implies social responsibility, many scientists use criteria of the public good in choosing and evaluating research. Who is to say they are wrong?

Finally, Merton's norms ignore the adversary nature of much scientific research. Science, I have argued, progresses in part through the conflict of competing views and traditions.[10] The assumption that scientists are guided by conflicting norms explains the frequent violations of the Mertonian ethos. A recent observer, Ian Mitroff, has proposed the following set of norms and counternorms, which are closer to the needs of science as I see them.[11] When different scientists observe these different norms in different situations, science achieves the creative conflicts and resolutions through which it advances. The opposing pairs of norms are:

faith in rationality	vs.	faith in rationality *and* irrationality (e.g., emotional commitment)
emotional neutrality	vs.	emotional commitment
universalism	vs.	particularism (e.g., highly creative people have special claims to discover knowledge)
individualism	vs.	societalism
community	vs.	solitariness (e.g., secrecy sometimes may be justified)
disinterestedness	vs.	interestedness (scientists are entitled to personal satisfaction and prestige)

293

impartiality	vs.	partiality (scientists must be concerned with the consequences of their discoveries)
suspension of judgment	vs.	exercise of judgment (judgments sometimes must be made on insufficient evidence)
absence of bias	vs.	presence of bias (scientists must acknowledge bias and take it into account)
professional loyalty	vs.	loyalty to humanity as a whole
freedom for research	vs.	management of research (science must be planned like any scarce national resource)

IS SCIENCE MORALLY NEUTRAL?

If I am correct, the scientist, acting according to implicit norms, makes moral choices again and again in the course of his research. He refuses to falsify data, refrains from "prettying" his conclusions to please his colleagues, tries to be fair to the work of colleagues he dislikes, and so on. But he generally does these things out of respect for truth, which in science is an intellectual and not a moral value. In science morality serves truth, not the reverse. Science, then, can be called morally neutral to the extent that it seeks to understand the world rather than improve it. Its aim is to discover truth, not to do good.

It may be objected that science makes a value choice when it seeks only public, objective truth and not other kinds. A Zen mystic, it may be said, or a Romantic poet would look very differently for the truth of nature. Quite so, but the preference for objective truth is an intellectual and not a moral one. Or is it? Until recently Western science refused to investigate such phenomena as telepathy, precognition, stellar influence, extraterrestrial civilizations, UFOs, and the conscious control of inner bodily processes by mystics and yogis. Was this not done from arrogance? Or intolerance? If so, how can science be morally neutral? (Telepathy? You expect us to take that seriously?) Now, to be sure, many scientists regard these phenomena as illusions, and until recently many probably regarded them as singularly fatuous. But the fundamental reason for lack of interest in these phenomena was the lack of instruments available for an adequate test of hypotheses. These areas were ignored partly from hard science chauvinism but mainly from the plausible belief that they could not yet be investigated objectively. The arrogance of certain scientists is not intrinsic to the scientific enterprise itself. New work may be resisted, but if it is testable, it will not be resisted indefinitely, for testability is the first requirement for any claim to objective knowledge. This requirement excludes certain claims from consideration, not certain phenomena. Scientific detachment is not arrogance.

Nonetheless, the claim that science always is morally neutral will not stand up to examination. In some areas the direction, though not the content, of research is guided by factors other than the aspiration for truth alone. The scientist can undertake only those projects for which money is available. Most money now is provided by government and industry for areas where it is hoped that research will have practical applications. Thus the external sponsors of science very often determine what research is undertaken, though not what conclusions are reached. This may well be desirable, depending on how generous the paymasters are in their perception of the human needs that science can meet; after all, a morally neutral science cannot be a humanitarian one. But it is well that we recognize the situation for what it is.

Moreover, science, as we have seen, always has been influenced by the world views and ideologies of the societies in which it is done, and to this extent it never has been morally neutral. World views and ideologies often express sectional and class interests—interests that may affect the scientist's choice of theoretical assumptions and research-guiding metaphors. Darwin, for instance, may have thought that he faced nature with an open mind. In fact he faced it with one that was partly that of his social class. He chose to explain natural phenomena by means of ideas drawn from laissez-faire economics, ideas which in turn expressed the desire of the middle class to pursue commerce without government regulation. In turning to these ideas, in finding them congenial, Darwin was influenced by the class interest he shared. Insofar as this interest included a belief in the moral desirability of certain social arrangements, Darwin's scientific thinking was not morally neutral. In Weimar Germany, again, many physicists abandoned determinist theories under the influence of a new climate of opinion mingling *Lebensphilosophie* and Spenglerian pessimism. The pressure on science of world views and ideologies is more subtle and pervasive than we have suspected.

My criticism of the moral neutrality thesis is mild, however, compared with the criticisms made in recent years by, among others, Jerome Ravetz, Jürgen Habermas, Herbert Marcuse, and Theodore Roszak.

Ravetz on the Corruption of Science. Present-day science, says Ravetz, is "industrialized."[12] Most research is financed by government and private industry in the hope of short- or long-term industrial spin-off. More importantly, science has changed its social character, for as research becomes impossible without a grant or contract, the scientific community devotes ever more time and energy to acquiring and distributing funds. According to Ravetz, a highly stratified professional society has grown up, consisting of funding agencies, a privileged group of scientists serving as advisers and contractors, a less privileged group of scientists working individually on small grants, a large number of Ph.D.'s and pre-Ph.D.'s employed on projects by contractors, and, finally, unemployed scientists with neither funds nor research. Science has become the core of a knowledge industry, converting scheduled inputs of manpower and money into scheduled outputs of applicable findings. Truth no longer is sought for its own sake. Science has ceased to liberate the mind from inadequate conceptions, and has ossified into a dogma.

Industrialized science, says Ravetz, leads to four kinds of abuse: "shoddy" science (pointless publication); "entrepreneurial" science (the quest for research grants to the neglect of disinterested research); "reckless" science (applied research activated by runaway technology); and "dirty" science (research projects with morally dubious aims and consequences).

Ivory tower, academic, morally neutral science has gone for good, declares Ravetz. Industrialized science is here to stay. However, if it is soundly led, this science can be made socially responsible. By solving technical problems set for it by government and industry, it can become a "service" science.

In addition, says Ravetz, we need a new, "critical" science, as different from contemporary science as that of the seventeenth century was from its medieval predecessor. This science will have new methods and a new ideology. It will serve humanity as a whole rather than any particular society, focus on the problems created by runaway technology, and seek to persuade social organizations to adopt the solutions it reaches. Ravetz's chief candidate for this role is ecologist Barry Commoner's Center for the Biology of Natural Systems at Washington University. The center does applied interdisciplinary work involving social as well as natural scientists. It has investigated important topics in agriculture and energy and has tackled such special problems as chemical carcinogens and nitrate pollution of rivers in the corn belt.

Is Ravetz right? His criticism of the moral and social state of contemporary science is devastating. But the critical science he proposes as a remedy is insubstantial. What he offers is more a declaration of hope that such a science will turn up than a description of what it will be like. We are told only that it will focus on the repair of technological damage and will be politically activist. But the problems of critical science are just those which in any case a socially responsible industrialized science is most likely to tackle. If so, what can critical science be but a pressure group? What in fact does the future hold? Not, to my mind, the replacement (by critical science) of academic basic science, absorbed into industrialized science, but rather the transformation of academic science into an enterprise that is more humanitarian and socially responsible. This science will be inspired by the quest for truth; but it also will relate this quest to other quests of men in society. It will be prepared not only to discern and promote the potentially beneficial consequences of its discoveries but also to avert the potentially injurious ones.

Ravetz now admits that his original conception of critical science was sketchy.[13] The idea occurred to him, he says, as he was seeking some way to overcome the opposition between academic and industrialized science. Recently, however, private individuals, using

297

publicly available information, have published effective critical studies of the misuses of science and technology. For Ravetz this is critical science. I call it freelance science. These individuals are doing science, certainly, but they have not created a science, and I find no evidence that they are likely to do so. There is no organized movement working with a theory or seeking one, no aim (except criticism of the status quo), no research tradition, no institutions, no peer reviews (Ravetz admits this last), in short no entity that can be called a science at all. In my view, Ravetz has mistaken a little scientific activity for a genuine scientific enterprise.

Habermas on the Hidden Interest of Science. Where Ravetz diagnoses the corruption of contemporary science, Jürgen Habermas sees a hidden interest at work ever since the seventeenth century—the interest of "technical control." [14] Men do science, he says, mainly because they wish to control nature. How does this interest govern scientific practice? The objects of technical control, says Habermas, must be recurrent, manipulable, measurable, and predictable. But the scientist too seeks to measure, manipulate, and predict. He therefore proposes laws summarizing the observed behavior of entities of the very kind that interest the technologist. When the technologist knows how objects will behave under certain conditions, he can select the right instruments to manipulate those conditions. So the technologist uses these laws, as the artisan uses the rules of his craft, as an aid to the control of objects. Modern science and technology thus are mutually adapted, serving the same interest.

Should we agree with Habermas? He is right to say that men have an interest in technical control, that scientific knowledge can be used to achieve such control, and that governments support science because of this interest. But science also serves, and always has served, an interest in pure understanding. This interest is aesthetic and contemplative, taking a delight in the harmony of nature. It also is emancipatory, since it frees man from fear and superstition and progressively enlarges his understanding. This was the aim of natural philosophy, and it is the aim of basic science today.

Habermas, then, argues from the true premise that scientific knowledge can serve an interest in technical control to the false conclusion that scientific inquiry has served this interest only. That this conclusion is false can be shown by a few historical examples. If Faraday said that one day the government would earn revenue from electricity, he also spent much of his career looking for the ultimate source of matter in fields of force. Ernest Rutherford, discoverer of the nucleus of the atom, dismissed the idea that splitting the atom would provide much energy. Einstein proposed his theory of special relativity to resolve an inconsistency between mechanics and electromagnetism, not because he foresaw that the equation $E = mc^2$ would have any practical application. Moreover, some sciences and theories seem incapable of aiding technical control. Darwin's theory of evolution yields no predictions, and the heavenly bodies cannot be controlled (though astronomical knowledge can be used to control other objects, as in agriculture and navigation).

Marcuse on Science's Dematerialization of Nature. Herbert Marcuse maintains that since the Renaissance science has served technology and hence the ruling class, which seeks continually to increase its control over men and nature.[15] Science has "dematerialized" nature, abstracting and idealizing those properties that can be handled mathematically. Science also has treated nature as a neutral, valueless stuff, to be manipulated ruthlessly as though it lacked life and goals of its own. Today, says Marcuse, science and technology are harnessed to a vast productive apparatus, which could do all the work now done by human brain and muscle. Instead, it is used to keep the masses working and hence docile, sating them with consumer goods to keep them from reflecting on their real servitude.

Marcuse calls for a new science and a new technology. The scientist, he declares, must study nature sympathetically as a living partner with values and potentialities of its own. He must seek knowledge which can be used to help nature realize its innate possibilities. The technology which such knowledge makes possible will cooperate with nature rather than control it. In liberating nature, man will liberate himself. In order to see the potential for life and development in nature, man must free his mind and senses from the many concepts which express the traditional Western idea that nature is a passive, inert object for human control. Self-liberation, social liberation, and the liberation of nature all imply one another.

Marcuse's notion of a new, sympathetic science strikes me as no more than a romantic hope. Modern science has made modern technology possible. What technology would a different science sustain? Marcuse does not say. But a "sympathetic" science would seem to imply a much less powerful technology and hence a return to a more primitive and laborious way of life. Marcuse insists that modern technology is necessary for human liberation because it can take over much of the repetitive work that men do now. Yet surely this technology can be provided only by the science he criticizes. Science, moreover, is the quest for the truth of nature, not for sympathy with nature. Marcuse is asking science to become more like poetry and mysticism, when the real need is for a new human being whose sympathetic and poetic side is developed along with his rational and scientific side. Marcuse asks science to change, when it really is culture that should change. Science and technology need to be seen as elements of our culture, not (with Marcuse) as its essence. Finally, Marcuse appears to identify science with physics, not realizing that something of the sympathetic attitude he advocates already is present in biology and the environmental sciences.

Other writers, such as Theodore Roszak, Henryk Skolimowski, and Brian Easlee,[16] have called for a "compassionate" science in which the scientist identifies to some extent with the phenomena he studies. Roszak criticizes science for claiming that abstract, mathematicized knowledge is the only true knowledge.

But scientists as a whole make no such claim. Only misguided dogmatists maintain that mathematical science is the sole source of truth. Easlee and Skolimowski want a compassionate science mainly, it seems, as an alternative to reductionist biology. They object to reductionism because it denies purposes to living things. But teleology and even vitalism are not incompatible with detachment. One can study phenomena dispassionately without devaluing them or denying their possible purposiveness. There can be intellectual respect as well as emotional sympathy.

In sum, I am not much impressed by any of these indictments.

300

The claim that science since the seventeeth century has been corrupted by a technological interest will not stand up to scrutiny. Nevertheless, science is not, and probably never has been, morally neutral. In the selection of topics to investigate, it is influenced by its paymasters, and in the choice of assumptions for its research traditions, it is moved by world views and ideologies.

SHOULD SCIENCE BE PLANNED?

Let us now look at the second moral issue raised by the interaction of science with the wider world. In this century, science has become an object of vital public concern, because for the first time in history scientific knowledge has become the basis of technology and hence a partner in the vast range of benefits and ills for which technology is responsible. Because scientific knowledge is technologically applicable, and therefore capable of both use and misuse, it no longer can be regarded as an unqualified good. So long as men were unable to use their knowledge of nature to harm one another or nature itself, the search for the truth about nature could be amply justified. But now that further scientific advances—in genetics, say, or in nuclear physics—can give men vast powers over nature or one another, it is an open question whether all branches of scientific inquiry should be pursued regardless of the consequences. Knowledge does not exist, nor is it always sought, in isolation from other human purposes. Hence, before we seek some type of knowledge, we first must ask how that knowledge is likely to be used.

How, then, can science be planned so that it promotes human welfare? What principles should be followed in selecting, from the vast array of possible problems, those most worthy of investigation? From the abundance of views on this issue, I have chosen four main approaches to the planning of science.

1. The case against planning has been stated powerfully by the chemist and philosopher *Michael Polanyi*, who compares the scientific community to the free market.[17] If each scientist is allowed to judge for himself where he can best contribute to current knowledge, an "invisible hand" will guide research into those areas where it is

301

most useful. Coordination comes about spontaneously as each scientist adjusts his work to the findings of others. In Polanyi's words, "the pursuit of science by independent self-coordinated initiatives assures the most efficient possible organization of scientific progress."[18] Scientists themselves should decide what money is to be spent on science, and they should do so by using purely scientific criteria. Thus Polanyi justifies freedom for science not as an inherent right of scientists but as a way to stimulate scientific growth.

A similar conclusion has been reached from different assumptions by *Derek de Solla Price*.[19] Whereas Polanyi is apt to regard science as a fragile organism vulnerable to outside interference, Price believes that science is such a hardy specimen that society cannot possibly hurt it. In Price's view, science develops worldwide according to its own laws, which no national plan can alter. Such a plan may act as a minor perturbation on science but cannot substantially deflect it. Stalin, for instance, stamped out Mendelian genetics in Russian but could not affect it in the West.

Jacob Bronowski offers another version of the antiplanning thesis. Where Polanyi responds to the Marxist claim that scientific growth is determined by economic forces, Bronowski reacts to the military use of science in Vietnam, Biafra, and the other battlegrounds of the 1960s.[20] According to him, science rests on individual integrity whereas modern government is corrupt. When the two meet, science is corrupted too. It is up to scientists "to set an incorruptible standard for public morality," to demonstrate "an unrelenting independence in the search for truth that pays no attention to received opinion or expediency or political advantage."[21] Hence science must be "disestablished" from the state. The government should make a single block grant for all research, which scientists would divide among themselves. Scientists, not civil servants or politicians, should weigh the social and scientific benefits of competing projects and should decide which projects to support.

What are we to make of these views? Price's position is refuted by the facts. As I have shown, science is persistently influenced by external factors. If Mendelian genetics flourished outside Stalinist Russia, the cause was not only the dynamism of scientific ideas themselves but also the willingness of governments and foundations to fund research in genetics, a science with important medical and agricultural applications.

Polanyi and Price ignore the consequences of scientific discovery. Suppose scientists, pursuing truth for its own sake, make a string of discoveries in nuclear physics or molecular biology that beg to be misused? And suppose they produce little knowledge relevant to urgent social problems such as overpopulation and pollution? Polanyi and Price seem to think that scientists can fiddle while the world burns. Bronowski is outraged, yet his solution is purely negative and to the same effect—science for its own sake. Moreover, Bronowski would also allow scientists the luxury of proving to themselves how much holier they are than other people.

Against Polanyi, Price, and Bronowski I maintain that scientists are not supremely competent to manage their own affairs. Granted, scientists themselves are best qualified to assess research findings, and the leaders of a field are likely to be the best judges of the research potential of particular persons and projects. But scientists are not the ideal persons to allocate funds among fields. This is not only because the public has an interest in the possible technological advantages of progress in one field rather than another but also because scientists are committed to their own fields and so are unlikely to choose impartially among them. Moreover, the scientific community is not qualified to lay down guidelines for socially applicable research, although it most certainly can and should advise on them.

2. According to a second view, proposed by *Alvin Weinberg*, Director of the Oak Ridge National Laboratory, Tennessee, science is a "technical overhead" on other social goals in the sense that particular scientific projects cannot be uniquely matched to specific social purposes.[22] Society, then, allocates to technology most of its resources for achieving social goals. In Weinberg's view, resources should be allocated instead to agencies responsible for particular social missions. These agencies should decide what proportion of their money to spend on scientific research. In judging what scientific fields and institutions to support, they should be guided partly by criteria internal to the field but mainly by external criteria. Internal criteria concern the quality of work done in a field—whether the field is ripe for development, for example, and whether its leaders are out-

standing scientists. Weinberg advances three criteria: How much does the field contribute to related fields? Are its findings of use to technology? Can they serve socially desirable ends? He then names some fields to which these criteria might be applied: molecular biology, high-energy physics, nuclear energy, manned space exploration, and the behavioral sciences. He himself favors nuclear energy and the behavioral sciences.

Yet, to my mind, technological agencies are likely to sacrifice the more esoteric fields to the claims of those with more practical applications. Indeed, the former fields would not be the responsibility of any particular agency. Many people, it is true, follow Polanyi in urging that each field make its case for funds as strongly as it can and exercise what has been called "proposal pressure" on possible sponsors.[23] But the pressure a field can muster is not a fair measure of the value of the projects it proposes, and, again, the more esoteric fields are likely to have fewer supporters and exercise less pressure.

3. A third view, put forward by *Joseph Ben-David*, treats science as a "social overhead investment."[24] Science, says Ben-David, is a social investment in that it contributes to society's purposes and productivity, and an overhead in that it contributes to all of them rather than to a single one. Hence, he argues, it should be paid for on the same grounds as other social services such as public health and education. Indeed, like education, science creates a pool of highly skilled manpower and basic knowledge which can be used for particular social purposes. However, Ben-David's view is open to the same criticism as Weinberg's. If science is an investment for future social productivity, what is the case for supporting research of scientific importance but no obvious social relevance? Such research may not be ignored but it surely will be slighted.

4. A fourth view, proposed by *Stephen Toulmin*, holds that as society satisfies its material needs and develops a high culture, it turns increasingly to spiritual and intellectual pursuits such as science.[25] In due course a "tertiary industry" of research, including natural science, emerges. As people are freed from the labor of producing goods, more and more of them will be employed in research. Indeed, says Toulmin, pure science already is being done in commercial research laboratories and already is "indispensable as a major source

304

of employment and prosperity."[26] However, attractive as this analysis is, it avoids the question whether science should proceed on its own course regardless of its technological consequences. Moreover, Toulmin does not propose criteria for assessing what part of a nation's resources should go to science, still less for deciding how much should go to particular sciences.

At present, then, neither scientists nor laymen can agree whether science should be regulated at all from outside, or what criteria should be used in allocating scarce resources to science. In my view, the community is entitled to participate in the choice of fields on which its money is spent. The objection that the general public and its representatives are likely to have little sympathy for research of no practical utility can be met in part by educating for a scientifically more literate citizenry. Creating institutional means for *some* democratic control of science will not be easy. It is just beginning to be undertaken. Nevertheless, the scientist no longer can ignore the moral and social implications of scientific research. How, then, should he take them into account?

THE SCIENTIST'S MORAL AND SOCIAL RESPONSIBILITY

I believe that every scientist has a specific, minimum moral and social responsibility. He should refrain from doing any research which may endanger the public or have technological uses potentially more harmful than beneficial. This responsibility takes precedence over his professional responsibility, his personal ambitions, and the advance of science itself. Who should see that it is met—the community of scientists or courts of law? I believe that the scientific community should first be allowed to regulate itself. If scientists fail, society must take over. Although the law looks favorably on experts who regulate themselves, those who fail consistently have had to accept laws more stringent than they themselves would have recommended.

Controls on Recombinant DNA. Let us consider a test case of cooperation between the scientific community and public authori-

305

ties—the regulation of research in recombinant DNA. During the late 1960s and early 1970s, research in gene manipulation and synthesis developed rapidly. In the spring of 1974 a powerful, new technique was created for cutting and recombining DNA molecules. Genes from one organism were linked to a vector molecule—either a piece of chromosomal DNA, known as a plasmid, or a virus such as a bacteriophage—and then implanted in another organism. The technique promises great benefits. Biologists should be able to transplant into bacteria the capacity to produce rare drugs such as insulin, and into crops the ability to fix nitrogen from the air. In human beings they should be able to substitute healthy genes for faulty ones to cure genetic diseases such as diabetes and genetic defects such as dwarfism.

But the technique also is dangerous. The addition to any organism of genes potentially harmful to man or other life forms could confer on it a selective advantage with disastrous results. Bacteria equipped with genes for drug resistance or for cancer- or toxin-formation might escape from the laboratory and swarm through human and other populations. Bacteria might develop new and uncontrollable characteristics. The biologist's favorite bacterium, *Escherichia coli*, has inhabited the human intestine probably since man first appeared on earth. Since most laboratory workers are exposed to the organisms they study, an *E. coli* might break out by this route and ravage the population at large.

Alarmed by these potential hazards, a committee led by Paul Berg at once urged biologists throughout the world to postpone two kinds of experiment and go cautiously with a third until an upcoming conference had debated the matter. Only two similar appeals have been made in recent memory. In the early 1940s non-German physicists called for a ban on publications in atomic research in order to deny the Axis powers strategic information. The research itself went ahead, however. In 1969 the Harvard team that isolated the first gene from a bacterium warned against government misuse of their discovery.

In February 1975 the conference (meeting at the Asilomar Center, in Pacific Grove, California) replaced the moratorium on certain types of experiment with general guidelines deferring some experiments and limiting others. Three categories of experiment with commensurate containment levels were adopted—low, moderate, and high. Containment could be physical, in the form of specially equipped laboratories, or biological, in the form of experimental organisms designed to self-destruct outside the laboratory.

Nevertheless, the conference almost failed to commit itself to guidelines at all. Many scientists regarded controls as high-minded intrusions on their freedom. James Watson declared that regulation was impractical; scientists would ignore it behind laboratory doors. Gene manipulators, he said, would have to live with the dangers of their work and realize that if they made mistakes, they would suffer the consequences. Joshua Lederberg maintained that guidelines might be turned into totally unacceptable laws by legislators suspicious of the field and unacquainted with its technicalities.[27] Moreover, since no accidents had occurred, conference members had no data by which to calculate the hazards precisely. Without data, they had to argue from moral principles—an activity to which scientists, as opposed to philosophers or lawyers, normally are unaccustomed. In the end, however, the scientists accepted their social responsibility.

A committee of leading biologists hammered the guidelines into more specific but still voluntary rules, and in June 1976 these were adopted with modifications by the National Institutes of Health (NIH). The rules represented a compromise between the fear of creating dangerous new hybrids and the fear of excessive regulation of research. Then the public, who should have been invited to participate earlier, entered the discussion. Communities across the nation debated whether to allow this research within their borders and whether the NIH rules were strong enough. In Cambridge, Massachusetts, the City Council voted a moratorium on research and then, after 7 months of heated debate, lifted it. This debate was unprecedented, and it led elsewhere to an unusual degree of citizen involvement in the conduct of research. Alarmed by the prospect of a patchwork of local regulations and voluntary guidelines, Congress began to debate the creation of compulsory federal laws. Passions rose among scientists. "Have we the right to counteract, irreversibly, the evolutionary wisdom of millions of years, in order to satisfy the ambition and curiosity of a few scientists?" asked Erwin Chargaff of Columbia. "History will curse us for it." Countered James Watson: "I told Sargent Shriver that recombinant DNA is the most overblown thing since his brother[-in-law] created the fall-out shelter debacle!"[28]

My concern here is not so much with the issues of the debate as with its influence on relations between scientists and the public.[29] At first scientists regarded recombinant DNA as a purely scientific problem to be solved within the scientific community itself. Researchers who conducted dangerous experiments could be brought into line through peer pressure and perhaps a cutoff of funds. There was no need to consult the public. The field of recombinant DNA was about to take off, and many biologists were impatient to get on with research. At scientific meetings there was more discussion of research than of its potential hazards. Pressure was put on the rule-drafters both to produce relatively lenient guidelines and to show the public that scientists could manage their own affairs.

Many scientists believed that if they made their own rules, the public would not intervene. How wrong they were. At Asilomar scientists adopted their own moratorium; at Cambridge the public imposed theirs. For four months a citizens' review board questioned a succession of scientists in order to decide for itself what the hazards were, how adequate the NIH guidelines were, and how far scientists were moved by self-interest. They made it plain that scientists are not entitled to have the last word on the moral and political implications of their work. The future of biomedical research, the setting of priorities, the weighing of the social costs and benefits of medical technology, these are political issues because they affect the lives of every citizen.

At Cambridge, then, the public won the right to some voice in the direction and conduct of scientific research. Recombinant DNA almost certainly is only the first in a series of issues in which scientists are required to justify their actions to the public's representatives. In the course of the controversy, scientists and the public have come to understand one another better.

In the seventeenth century Bacon insisted on science's responsibility to humanity. Knowledge, he said, confers power; therefore, seek knowledge that can be used well. The Royal Society, seeking to realize his ideal, promised that it would benefit mankind. Bacon's dream of an alliance between science and technology was not realized until the late nineteenth century. But when the time came for science to assume its social responsbility, scientists claimed the right to seek knowledge regardless of the consequences. All knowledge is good, they said; it is you who misuse it, not we. No gods themselves, they expected men to behave with a god's restraint.

This attitude has changed in the last five years. The end of easy money has made scientists realize that the public is under no obligation to support them. On the contrary, it is they who are obligated to the society which pays for their research and reaps its consequences.

It may be asked whether the general public can understand science well enough to make useful judgments about it. Can representatives of the public assess the potential benefits and dangers of a particular project? Are they entitled to set priorities in a publicly funded field? I reply that although they cannot do the experiments, members of the public can comprehend at least the basic elements of the science involved. They also can appraise the potential social consequences of research as farsightedly as scientists themselves. They therefore should be allowed an equal, and perhaps even a deciding, voice in assigning priorities among projects for which they are paying and whose consequences may well affect them.

Public participation is urgently needed in the case of recombinant DNA, whose potential risks and benefits exceed those of harnessing the atom. DNA technology promises to give man a power over nature that is at once more creative and more dangerous than any he has acquired so far. This is the power to design new organisms on the spot rather than await the slow, random reshuffling of genes that occurs in nature. Hitherto evolution has seemed as irrevocable as entropy or time. Now man has a hand on the force that made him. Or is the hand held by the force? Is the human race in its present form due to be superseded by a mutant species which it has itself created? And will that species be superior to man or inferior? These questions challenge not only science but human wisdom itself.

309

Alternatives. Some democratic control of science is desirable
and inevitable. What can scientists do to bring about a harmonious
arrangement? Above all, they can educate the general public to understand the nature of science and participate in its government.
They can do so best by openly discussing the aims and limitations of
the scientific enterprise. The scientist can and should explain his
work in ways the public can understand. All the media are open to
him: radio, television, publications of every kind. As a scientist, he
usually is assured of a hearing. He also should try to counteract mistakes and misrepresentations made by the media working in their own
interest. New drugs often are credited with powers they do not
possess, and many discoveries are overdramatized.[30] Only a scientifically informed public can debate the manifold extensions of
science and technology into everyday life without giving way to shal-
low optimism or frenetic hostility.

Scientists also should take a greater interest in the teaching of
science, discuss their work with teachers, and explain it personally to
students. Young people can understand the nature of science better if
a scientist encourages them to do research on their own, posing and
solving problems within the limits of their experience and understanding. The scientist also should present science as the human,
fallible undertaking it is. He should encourage the teacher to cul-
tivate a critical, questioning spirit in the young. This should be
regarded as a virtue, not a luxury.

Finally, the scientist should form or join organizations dedicated to influencing governments, corporations, and other agencies
that use scientific knowledge. In recent years scientists have taken a
much more active part in either supporting or opposing different
policies and projects involving applied science and technology, such
as the ABM system, the supersonic jetliner, and nuclear power stations. Alternatively, the scientist can put his knowledge at the service
of lay groups working for specific political and social goals. This
much is sure: the social consequences of science become more important every year, and it is up to all of us to deal with them more
effectively.

SUMMARY

In this chapter I have examined two issues raised by the interaction of science and the wider world: Is science morally neutral? and Can science be planned? In their quest for the truth of nature, scientists observe a variety of largely tacit and apparently conflicting norms which never have been codified satisfactorily. The most ambitious attempt to list them—Merton's—must be judged too highly idealized, since the norms he proposed are frequently violated and they do not sufficiently respect the adversary nature of the scientific enterprise.

It often has been claimed that since truth is an intellectual value, science itself is morally neutral. In my view, this claim is mistaken, first because the broad direction of much research is determined by external funding agencies, and second because scientists, some more than others, are affected by the world views and ideologies of their time. More radical criticisms have been made of the alleged neutrality of science. Ravetz maintains that contemporary science has been corrupted by easy money. But the critical science he proposes as an alternative is a marginal activity. Habermas declares that science has been guided since the seventeenth century by an interest in the technological control of nature. But he overstates his case, apparently assuming that because science can be guided by this interest, it consistently has been. Marcuse, Roszak, and others call on scientists and technologists to sympathize and cooperate with nature. But they should blame society more than science for the plunder of nature, since it is mostly laymen who put scientific knowledge to work.

Since society pays for, and is affected by, scientific research, science should produce knowledge that is socially useful. Just how much research should be guided by this principle is open to debate. Some thinkers, such as Polanyi and Bronowski, object to the planning of science. But they ignore society's right to useful scientific knowledge; they overestimate the ability of scientists to distribute funds wisely among fields; and they play down the very real possibility that scientists can pursue goals that are dangerous to the public. Weinberg and Ben-David, on the other hand, provide for the interests of technology but overlook those of pure science.

311

As this disagreement between experts shows, it will not be easy to create institutions for the democratic control of science. Meantime, how is the scientist to meet his responsibility to society? Mainly, I suggest, by refraining from research which may endanger the public or have technological consequences that on balance are harmful. Molecular biologists have acknowledged this principle in part, by adopting voluntary guidelines for the conduct of recombinant DNA research. The public also has asserted its interest in the principle by holding hearings on gene manipulation, and the federal government is certain to pass laws regulating this field. The recombinant DNA controversy is likely to be the first of many debates in which scientists and the public come to appreciate better their mutual rights and responsibilities.[31] If this book helps to further such a rapprochement, it will have served its purpose.

NOTES

1. I do not wish to imply that truth is the only value sought by science. Beauty in the form of simplicity is another. Thus Einstein wrote: "The grand aim of all science . . . is to cover the greatest possible number of empirical facts by logical deductions from the smallest possible number of hypotheses or axioms." Albert Einstein, *Ideas and Opinions*, trans. and rev. Sonja Bargmann (New York: Crown Publishers, 1954), p. 282. On the value of beauty, see Paul Dirac, "The Evolution of the Physicist's Picture of Nature," *Scientific American* 208 (1963), 47: "It seems that if one is working from the point of view of getting beauty in one's equations, and if one has a really sound insight, one is on a sure line of progress. If there is not complete agreement between the results of one's work and experiment, one should not allow oneself to be too discouraged, because the discrepancy may well be due to minor features that are not properly taken into account and that will get cleared up with further developments of the theory. That is how quantum mechanics was discovered."

2. See Martin Gardner, "Great Fakes of Science," *Esquire* 88 (October 1977): 88 ff. See also Arthur Koestler, *The Case of the Midwife Toad* (New York: Random House, 1971).

3. "Development of the Physicist's Conception of Nature," in *The Physicist's Conception of Nature*, ed. Jagdish Mehra (Dordrecht, Holland, and Boston: Reidel, 1973), pp. 13–14.

4. Jacob Bronowski, *Science and Human Values*, pp. 58, 68.

5. *Social Theory and Social Structure* (New York: Free Press, 1957), pp. 552–53.

6. E.g., Bernard Barber, *Science and the Social Order* (New York: Collier, 1962); and Norman W. Storer, *The Social System of Science* (New York: Holt, Rinehart, and Winston, 1966).

7. Max Planck, *Scientific Autobiography and Other Papers*, trans. F. Gaynor (New York: Philosophical Library, 1949), pp. 33–36.

8. See S. B. Barnes and R. G. A. Dolby, "The Scientific Ethos: A Deviant Viewpoint," *Archives Européennes de Sociologie* 11 (1970): 3–25; and Michael Mulkay, "Aspects of Cultural Growth in the Natural Sciences," *Social Research* 36 (Spring 1969): 22–52.

9. Stewart S. West, "The Ideology of Academic Scientists," *I.R.E.* [Institute of Radio Engineers], *Transactions on Engineering Management* (June 1960): 61.

10. Merton has revised his views several times and now admits that there are conflicting norms in science. However, he treats them as "major norms and . . . minor counter-norms" rather than as norms of equal standing, and he has not used this insight to alter his original thesis.

11. Ian I. Mitroff, *The Subjective Side of Science*, p. 79.

12. Jerome R. Ravetz, *Scientific Knowledge and Its Social Problems.*

13. Address to the American Association for the Advancement of Science, 143d Annual Meeting, Denver, Colo., Feb. 21, 1977.

14. Jürgen Habermas, *Knowledge and Human Interests* (Boston: Beacon, 1971); and *Towards a Rational Society* (Boston: Beacon, 1970).

15. Herbert Marcuse, *One Dimensional Man* (Boston: Beacon, 1964); and *Counterrevolution and Revolt* (Boston: Beacon, 1973).

16. Theodore Roszak, *The Making of a Counter Culture* (Garden City, N.Y.: Doubleday, Anchor Books, 1969); Brian Easlee, *Liberation and the Aims of Science: An Essay on the Obstacles to the Building of a Beautiful World* (London: Chatto & Windus for Sussex University Press, 1973); Henryk Skolimowski, "Problems of Rationality in Biology," in *Studies in the Philosophy of Biology: Reduction and Related Problems*, ed. Francisco J. Ayala and Theodosius Dobzhansky (Berkeley and Los Angeles: University of California Press, 1974), pp. 203–24.

17. "The Republic of Science," in *Criteria for Scientific Development*, ed. Edward Shils, pp. 2–20.

18. *Ibid.*, p. 3.

19. "The Science of Scientists," *Medical Opinion Review* 10 (1966): 88–97.

20. Jacob Bronowski, "The Disestablishment of Science," in *The Biological Revolution: Social Good or Social Evil?* ed. Watson Fuller (Garden City, N.Y.: Doubleday, Anchor Books, 1972), pp. 305–21.

21. *Ibid.*, p. 313.

22. Alvin M. Weinberg, "Criteria for Scientific Choice," in *Reflections on Big Science* (Cambridge: MIT Press, 1967).

23. Harvey Brooks, *The Government of Science* (Cambridge: MIT Press, 1968), pp. 76–77.

24. Joseph Ben-David, *Fundamental Research and the Universities* (Paris: Organization for Economic Cooperation and Development, 1968).

25. Stephen Toulmin, "The Complexity of Scientific Choice II: Culture, Overheads or Tertiary Industry?" in *Criteria for Scientific Development*, ed. Shils, pp. 119–33. A similar view is advanced by Harry Johnson, "Federal Support of Basic Research: Some Economic Issues," *Minerva* 3 (1965): 500 ff. Science, for Johnson, is one of the highest expressions of a civilized society and deserves support for its own sake like art or music. However, Johnson regards science as a luxury item rather than (like Toulmin) as an activity satisfying a basic social need.

26. Toulmin, "Complexity of Scientific Choice II," p. 133.

27. *Science News* 107 (March 22, 1975): 195.

28. *Science* 194 (November 12, 1976): 705.

29. For a summary of the issues, see Clifford Grobstein, "The Recombinant-DNA Debate," *Scientific American* 237 (July 1977): 22–33. Grobstein considers the potential biohazards of the gene-splicing method and the effectiveness of plans for containing them.

30. In *The Visible Scientists* (Boston: Little, Brown, 1977) Rae Goodell maintains that the media have attracted a few scientists with colorful personalities and a flair for publicity. These "visible" scientists (e.g., Paul Ehrlich, Linus Pauling, Margaret Mead, William Shockley), she says, have come to represent science to the general public. I reply that the media have a responsibility to invite leading scientists to speak or write for them, no matter how little charisma they may have. Television, for example, can maintain audience interest in a dull speaker by providing an interesting background.

31. The last major survey of public understanding and appreciation of science took place over two decades ago. Recent data, says Clyde Z. Nunn, are "meager." They

reveal that (a) the majority of Americans show "considerable interest" in science; (b) their attitude is one of "ambivalence, not rejection"; (c) they want scientists to become more interested in problems of health, environment, energy, and social welfare; and (d) they recommend that the news media be used more extensively to communicate scientific information. ("Only 11 percent of the nation's daily newspapers have science editors.") See Clyde Z. Nunn, "Is there a Crisis of Confidence in Science?" *Science* 198 (9 December 1977): 995.

Selected Bibliography

Achinstein, Peter. *Concepts of Science: A Philosophical Analysis*. Baltimore: Johns Hopkins University Press, 1968.

Ben-David, Joseph. *The Scientist's Role in Society: A Comparative Study*. Englewood Cliffs, N.J.: Prentice-Hall, 1971.

Bernal, John D. *Science in History*. 4 vols. Cambridge: M.I.T. Press, 1971.

Beveridge, W. I. B. *The Art of Scientific Investigation*. New York: Norton, 1957.

Bronowski, Jacob. *Science and Human Values*, rev. ed. New York: Harper & Row, 1965.

—— *The Ascent of Man*. Boston: Little, Brown, 1974.

Bunge, Mario. *Scientific Research*. 2 vols. New York: Springer, 1967.

Campbell, Norman. *What Is Science?* 1921. Reprint. New York: Dover, 1953.

Cardwell, D. S. L. *Turning Points in Western Technology: A Study of Technology, Science, and History*. New York: Neale Watson, 1972.

Caws, Peter. *The Philosophy of Science: A Systematic Account*. Princeton, N.J.: Van Nostrand, 1965.

Commoner, Barry. *The Closing Circle: Nature, Man, and Technology*. New York: Knopf, 1971.

Crane, Diana. *Invisible Colleges: Diffusion of Knowledge in Scientific Communities*. Chicago: University of Chicago Press, 1972.

Einstein, Albert. *Out of My Later Years*. New York: Philosophical Library, 1950.

Feyerabend, Paul K. *Against Method*. London: New Left Books, 1975.

Greenberg, Daniel S. *The Politics of Pure Science*. New York: New American Library, 1968.

Greene, Marjorie, and Everett Mendelsohn, eds. *Topics in the Philosophy of Biology*, Boston Studies in the Philosophy of Science. Vol. 27. Dordrecht, Holland, and Boston: Reidel, 1976.

Gruber, Howard E. *Darwin on Man: A Psychological Study of Scientific Creativity, Together with Darwin's Early and Unpublished*

Notebooks, transcribed and annotated by Paul H. Barrett. New York: Dutton, 1974.

Hagstrom, Warren O. *The Scientific Community*. New York: Basic Books, 1965.

Hanson, Norwood Russell. *Patterns of Discovery: An Inquiry into the Conceptual Foundations of Science*. Cambridge: Cambridge University Press, 1958.

—— *The Concept of the Positron: A Philosophical Analysis*. Cambridge: Cambridge University Press, 1963.

Harré, Romano M. *The Principles of Scientific Thinking*. London: Macmillan, 1970.

Hempel, Carl G. *Aspects of Scientific Explanation and Other Essays in the Philosophy of Science*. New York: Free Press, 1965.

Holton, Gerald. *Thematic Origins of Scientific Thought: Kepler to Einstein*. Cambridge: Harvard University Press, 1973.

Howson, Colin, ed. *Method and Appraisal in the Physical Sciences: The Critical Background to Modern Science, 1800–1905*. Cambridge: Cambridge University Press, 1976.

Jonas, Hans. *Philosophical Essays: From Ancient Creed to Technological Man*. Englewood Cliffs, N.J.: Prentice-Hall, 1974.

Kearney, Hugh F. *Science and Change, 1500–1700*. New York: McGraw-Hill, 1971.

Kemeny, John G. *A Philosopher Looks at Science*. Princeton, N.J.: Van Nostrand, 1959.

Kuhn, Thomas S. *The Structure of Scientific Revolutions*. 2nd rev. ed. Chicago: University of Chicago Press, 1970.

Lakatos, Imre, and Alan Musgrave, eds. *Criticism and the Growth of Knowledge, Proceedings of the International Colloquium in the Philosophy of Science, 1965*. Cambridge: Cambridge University Press, 1970.

Laudan, Larry. *Progress and Its Problems: Toward a Theory of Scientific Growth*. Berkeley and Los Angeles: University of California Press, 1977.

Lloyd, G. E. R. *Early Greek Science: Thales to Aristotle*. New York: Norton, 1970.

—— *Greek Science after Aristotle*. New York: Norton, 1973.

Manuel, Frank E. *A Portrait of Isaac Newton*. Cambridge: Harvard University Press, 1968.

Maslow, Abraham H. *The Psychology of Science: A Reconnaissance*. New York: Harper & Row, 1966.

318

Mason, Stephen F. *A History of the Sciences*. New York: Collier, 1962.

Maxwell, Nicholas. "The Rationality of Scientific Discovery, Part I: The Traditional Rationality Problem." *Philosophy of Science* 41 (June 1974): 123–53.

—— "The Rationality of Scientific Discovery, Part II: An Aim-Oriented Theory of Scientific Discovery." *Philosophy of Science* 41 (September 1974): 247–95.

McCormmach, Russell, ed. *Historical Studies in the Physical Sciences*. Vols. 1–3, Philadelphia: University of Pennsylvania Press, 1969–71. Vols. 4–7, Princeton, N.J.: Princeton University Press, 1975–76.

Medawar, Peter Brian. *Induction and Intuition in Scientific Thought*, Jayne Lectures for 1968. Philadelphia: American Philosophical Society, 1969.

Merton, Robert K. *The Sociology of Science: Theoretical and Empirical Investigations*. Edited with an introduction by Norman W. Storer. Chicago: University of Chicago Press, 1973.

Mesthene, Emmanuel. *Technological Change: Its Impact on Man and Society*. Cambridge: Harvard University Press, 1971.

Mitroff, Ian I. *The Subjective Side of Science: A Philosophical Inquiry into the Psychology of the Apollo Moon Scientists*. New York: Elsevier, 1974.

Mumford, Lewis. *Technics and Civilization*. New York: Harcourt and Brace, 1934.

Nagel, Ernest. *The Structure of Science: Problems in the Logic of Scientific Explanation*. New York: Harcourt, Brace, and World, 1961.

National Research Council, Committee in Support of Research in the Mathematical Sciences. *The Mathematical Sciences*. Washington, D.C.: National Academy of Sciences, 1968.

Needham, Joseph. *Science and Civilization in China*. 4 vols. to date. Cambridge: Cambridge University Press, 1954–.

Polanyi, Michael. *Personal Knowledge: Towards a Post-Critical Philosophy*. Chicago: University of Chicago Press, 1958.

Popper, Karl R. *Conjectures and Refutations: The Growth of Scientific Knowledge*. 4th rev. ed. London: Routledge & Kegan Paul, 1972.

—— *The Logic of Scientific Discovery*. New York: Harper & Row, 1968.

Rapp, Friedrich, ed. *Contributions to a Philosophy of Technology*. Dordrecht, Holland: Reidel, 1974.

Ravetz, Jerome R. *Scientific Knowledge and Its Social Problems*. New York: Oxford University Press, 1971.

Shils, Edward Albert, ed. *Criteria for Scientific Development: Public Policy and National Goals,* a selection of articles from *Minerva.* Cambridge: M.I.T. Press, 1968.

Suppe, Frederick, ed. and introd. *The Structure of Scientific Theories.* Urbana: University of Illinois Press, 1974.

Teich, Mikuláš, and Robert Young, eds. *Changing Perspectives in the History of Science: Essays in Honor of Joseph Needham.* London: Heinemann, 1973.

Thackray, Arnold. *Atoms and Powers: An Essay on Newtonian Matter-Theory and the Development of Chemistry.* Cambridge: Harvard University Press, 1970.

Watson, James D. *The Double Helix: A Personal Account of the Discovery of the Structure of DNA.* New York: Atheneum, 1968.

Williams, J. Pearce. *Michael Faraday: A Biography.* London: Chapman and Hall, 1965.

Ziman, John M. *Public Knowledge: An Essay Concerning the Social Dimension of Science.* London: Cambridge University Press, 1968.

—— *The Force of Knowledge: The Scientific Dimension of Society.* Cambridge: Cambridge University Press, 1976.

Index

Index

Index

Index

Index

Index

Index

Index

Index

Index

332

Index